PIONEERS OF ONENESS

THE SCIENCE AND SPIRITUALITY OF UFOs AND THE SPACE BROTHERS

Gerard Aartsen

Pioneers of Oneness –
The science and spirituality of UFOs and the Space Brothers

First published October 2020. (Revised August 2021)

Copyright © 2020, Gerard Aartsen.

ISBN-13/EAN-13: 978-90-830336-00 (hardcover edition)

Published by BGA Publications, Amsterdam, the Netherlands.

www.bgapublications.nl

Typeset in Calisto MT and Calibri.

Cover design: Sorina Constantin
Cover photo: Michael Schneider

This book is dedicated to

Helena P. Blavatsky (1831-1891)
and
George Adamski (1891-1965),

without whose pioneering work and bravery
it could not have been written.

Contents

Acknowledgements
The author wishes to express his gratitude to Meryl
Tihanyi and Marc Gregory for their help in preparing
the material for publication. He is also very grateful to
Michael Schneider for his kind permission to use his
photograph 'Halo at Arosa' for the front cover, and Sorina
Constantin for her striking cover design. As always, the
author is deeply indebted to the late Benjamin Creme for
his inspiration and writings.

INTRODUCTION

When future generations look back at this time in history, the 1950s will be seen as the decade 'when it all started'.

Since the 1950s the world has been faced with the phenomenon that is now officially termed 'Unidentified Aerial Phenomena' (UAP), formerly known as Unidentified Flying Objects (UFOs). Although reports of the phenomenon can be traced back to mankind's earliest written records, as documented by Lord Desmond Leslie in 1953, and first began to attract the attention in modern times with the sighting of 'foo fighters' on the tail of World War II fighter planes, it only burst onto the world stage of mass popular interest with the reported sightings and landings of flying saucers in the 1950s.

Also in the 1950s, large sections of humanity were subjected to the Cold War of ideologies – the 'freedom' of capitalism and the 'justice' of communism – that was fought under the threat of atom and hydrogen bombs, the latest inventions of mass lethal force to spring from the fountain of human ingenuity.

Amid the initial furore over the news that life on

planet Earth is not an isolated incident in the cosmos when visitors from other planets landed in space craft in the 1950s, their message about the oneness of life, and the warnings against the threat of our nuclear self-annihilation were soon drowned in confusion caused by official denials and cover-ups, triggering ridicule and a growing chorus of demands for hard evidence. As a result, a debate erupted even in the late 1950s about the nature of the flying saucers which, without solid evidence, should more aptly be referred to as 'Unidentified Flying Objects'.

Fast forward to 2020 and the confusion is 'completer' than ever, with self-styled experts stumbling over each other like frenzied customers in a Black Friday department store sale – not because they know what 'Unidentified Aerial Phenomena' are, but because they came up with yet another theory for what has by now been reported by millions of people around the world, continues to baffle pilots and civilian eyewitnesses alike, and remains as elusive as it became when personal encounters with extraterrestrial visitors were 'officially' ruled out of court by the nuts-and-bolts Ufologists.

If we learn anything from these developments it should be that that when we look at phenomena in isolation, it is impossible to see the bigger picture. Clearly, the UFO phenomenon should be approached as a complex aspect of reality, and as long as it is considered separately from context, history, experiences, or correspondences in other fields – merely based on empirical data collected from sightings and the many theories on offer that equally lack connection with the larger system within which it occurs – it will continue to defy explanation or understanding.

Originating in the 1950s, not without significance in this context, systems theory has helped modern science to look beyond the boundaries of its separate disciplines, by approaching reality as an interconnected and interrelated system. As it looks for similarities among different systems, from simple to complex, systems science combines the findings from different disciplines into a hypothesis that finds the broadest support in the widest range of evidence.

Building on systems science, and taking consciousness as a fundamental constituent of reality, in order to establish the scientific value of people's spiritual experiences the Scientific and Medical Network proposes as methodological keys that, "the experience is recordable or reportable, that it is communicable and to some extent shared or potentially intersubjectively available."*

These certainly apply to the many accounts of the 1950s contactees. Almost without exception the people who reported being contacted by visitors from space in the era of initial contact, and many since, described it as a profoundly spiritual experience, and many people have said that even the sighting of a space craft changed their perspective on life forever.

In order to decide if personal experiences can be trusted to reflect some level of truth beyond the strictly subjective, in *Pioneers of Oneness* I have broadened my own cross-disciplinary approach to include not only the accounts of the contactees in general, and George Adamski's in particular, alongside the wisdom teachings and religious

---

* Harald Walach (2019), *Beyond a Materialist Worldview. Towards an Expanded Science*, p.75

postulates, but the latest insights derived from systems science and consciousness research as well, against the backdrop of the historical and social developments of our time.

After exploring the means and methods for reliable and meaningful research in Chapter 1, we examine the evidence that emerges from the correspondences and concurrences found in the wide range of sources and disciplines considered here, as we inquire into the reality and nature of extraterrestrial visitations from the perspectives of the nature of reality itself, evolution, technology, and time. After all, in our search for a broader understanding of reality we need to let the facts speak before we allow our present convictions to reject them. That is the method of true science – and the beginning of wisdom.

Taking the most fundamental approach to UFO research yet, *Pioneers of Oneness* is the result of my search to reconcile the basic tenets found in the contactees' accounts, the wisdom teachings, and the forefront of science. The vista that presents itself from this broader view of life and reality amid the turmoil of the present world crises is one of scientific and spiritual regeneration for planet Earth, as it takes its rightful place among the planets and peoples that populate the solar system.

Amsterdam, September 2020

"Humanity is now too clever to survive without wisdom."
–Ernst F. Schumacher, statistician and economist

1. UFO RESEARCH: HOW NUTS AND BOLTS GET IN THE WAY OF REALITY

Although the question of life beyond our planet has long engaged science, serious researchers do not usually refer to anything more specific than 'extraterrestrial life', to avoid the impression they expect to find much more than traces of micro-organisms. So, when the Massachusetts Institute of Technology (MIT) in April 2020 published a roundup of the latest scientific findings in the search for life outside Earth, its title revealed more than the modest volume suggests at first sight.

In order to appreciate this revelation, it must be remembered that the first newspaper report of a man claiming to have met visitors from space appeared on 24 November 1952 in the *Phoenix Gazette* (Arizona), with many more such reports in the years that followed – in America and the rest of the world. Of course, these stories were soon debunked and while their originators enjoyed their moment of fame, they have since been relegated to the fringes of Ufology by those who knew how to conduct proper research based on verifiable data.

However, without the accounts of the 'contactees' of

the 1950s the notion of human-like visitors from space stepping out of their craft and contacting people on Earth would probably not have been normalised or popularised to the extent that, 68 years of derision and ridicule later, an MIT publication on such an elusive phenomenon as life outside Earth would be unapologetically titled *Extraterrestrials*.

To be sure, as far as flying saucers or UFOs are concerned, "what people sometimes don't get about science is that we often have phenomena that remain unexplained," according to MIT astrophysicist Sara Seager in the *New York Times* in 2017.[1] Yet, it is precisely this attitude towards the phenomenon that was challenged about a decade earlier by two professors citing the "production of (un) knowledge about UFOs", mostly conducted for the purpose of "ignoring UFOs, constituting them as objects only of ridicule and scorn". In their 2008 paper 'Sovereignty and the UFO', International Security professor Alexander Wendt and Political Science professor Raymond Duvall went so far as to conclude there is "a prohibition in the authoritative public sphere on taking UFOs seriously, or 'thou shalt not try very hard to find out what UFOs are'."[2] The biggest mystery, according to their paper, is why both scientists and governments consider UFOs "not an 'object' at all, but a *non*-object, something not just unidentified but unseen and thus ignored".

They were not the first to criticize government efforts. The infamous Project Blue Book was pointedly called "The Society for the Explanation of the Uninvestigated"[3] when the US Air Force, after investigating over 12,000 reports of UFO sightings between 1952 and 1969, concluded that there was no evidence that these represented anything out of the ordinary, let alone that they might be extraterrestrial craft. Referring to the methods used, astronomer J. Allan Hynek, who acted as a scientific advisor to the project, called one of its members "the master of the possible: possible balloon, possible aircraft, possible birds, which then became, by his own hand (and I argued with him violently at times) the probable"[4], meaning that other avenues of research were deliberately left unexplored.

A string of recent newspaper reports, however, seems to indicate that the time of blanket government denial may finally have come to an end. For instance, in December 2017 the *New York Times* revealed that, despite repeated denial of any official interest in UFOs, the US Department of Defense had conducted a cataloguing of sightings recorded by military pilots from 2007 until 2012, when government funding dried up, as part of the Advanced Aerospace Threat Identification Program (AATIP).[5]

The *Times* report signalled a series of unusual official 'leaks' and reports that indicate a growing acknowledgement of government interest in what are now referred to as 'unidentified aerial phenomena', or UAPs, beginning in August 2017 with the declassification by the US Department of Defense of videos of UAP encounters by US Navy fighter jets based on the aircraft carrier USS Nimitz, off the coast of San Diego, that were recorded between 2004 and 2014. And in May 2019, US Navy pilots told the *Times* that from the summer of 2014 until March 2015 they had reported observing objects, this time on the US east coast, that had no visible engine or infrared exhaust plumes and could reach altitudes of 10,000 metres, and hypersonic speeds. Pilots Lt. Ryan Graves and Lt. Danny Accoin said they and three other Navy pilots first spotted the anomalous objects after their outdated radar system had been upgraded, but since 2014 they began to see the objects with the naked eye and recorded them on video with onboard cameras.[6]

Even more remarkable is the account of Joe Montaldo, who earned three degrees in Advanced Electronics and Avionics during his time in the US Navy. In a recent interview with a guest on the *UFO Undercover* podcast that he hosts, Mr Montaldo stated that the sighting from 2014 reminded him of something he had seen while he was stationed on the USS Nimitz himself. "We used to see these things that would come by the carriers, and they got good footage of these. They called them flat discs… They would come and would actually fly deck level, zip by, zip by. (…) And they got one of them that actually hovered on the edge, pretty sure it sat down on the landing, I mean the take-off end of the carrier. And I know they got video footage of it."[7]

In response to a Freedom of Information Act inquiry the US Navy in September 2019 for the first time confirmed that it "considers the phenomena contained/depicted in those 3 [declassified] videos as unidentified"[8] and on 27 April 2020 the Pentagon officially released the three videos that had been leaked since 2007.[9] On 23 July 2020 the *Times* revealed that the programme, renamed as Unidentified Aerial Phenomenon Task Force, was continued after 2012 as part of the Office of Naval Intelligence[10] and on 14 August CNN reported that, according to two defence officials, the Pentagon "is forming a new task force to investigate UFOs that have been observed by US military aircraft".[11]

In their paper professors Wendt and Duvall, who state explicitly they do not presuppose an extraterrestrial origin of UFOs, say "our puzzle is not the familiar question of ufology, 'What are UFOs?' but, 'Why are they dismissed by the authorities?' Why is human ignorance not only unacknowledged, but so emphatically denied? In short, why a *taboo?*" They assert: "The question today is not 'Are UFOs ETs?' but 'Is there enough evidence they *might* be to warrant systematic study?' By demanding proof of ETs first, skeptics foreclose the question altogether."[12]

Indeed, this persistent lack of transparency regarding the reality of unidentified objects or extraterrestrial visitors continued for many decades, with the likely ulterior motive of influencing the public's perception of their true nature through controlled leaking of sparse information, probably mixed with misinformation. But the fact that the US Navy now allows sightings to be discussed, declassified the videos, and acknowledged that it does investigate sightings by military personnel, has resulted in a subject that was long considered too 'fringe' now filling the columns of the *New York Times* and *The Washington Post* with increasing regularity.

Another development that indicates the UFO taboo may finally be broken is the founding of the Five Continents UFO Forum, meant to bring about UFO disclosure through the United Nations. An earlier attempt in 1978 by then-Prime Minister of Grenada Eric Gairy to put the subject of UFOs on the UN's agenda was abandoned after the Prime Minister had been overthrown in a coup.[13] The formation of the new international UFO research organisation was announced at a meeting of the Chinese UFO Association (CUA) in Chongqing at the weekend of 30 July 2018. The Five Continents Forum's preliminary meeting was held from 16-17 October 2018 at the Cosmos Hotel in Moscow and was reportedly funded by the Chinese government, with scholars and experts from 35 countries across five continents attending to discuss advancements in aeronautics, astronautics and extraterrestrial fields. The Forum's headquarters will be based in Heze City, in Shandong Province.[14]

As government and scientific research into unidentified aerial phenomena and extraterrestrial life is moving from denial towards a growing willingness to finally explore possibilities and hypotheses, there is a steady rise in 'serious' research in these areas.

On the one hand, we see research that focuses exclusively on what the current state of material science and technology allows for, such as summarized in the MIT book *Extraterrestrials*. Here, science journalist Wade Roush provides an overview of "the question itself, why it remains unanswered, and how scientists are trying to answer it". For instance, he describes an updated version of the Drake Equation, to estimate the number of potential inhabited worlds in the Milky Way, not based on the likely number of communicative civilizations but rather on the list of potential detectable biosignature chemicals in the atmospheres of inhabited worlds.[15]

In a related effort, whose findings were reported in *The Astrophysical Journal* of 15 June 2020, scientists at the University of Nottingham found that there should be at least 36 Communicating Extra-Terrestrial Intelligent (CETI) civilizations within our galaxy, based on assumptions that derive from "the one situation in which intelligent, communicative life is known to exist – on our own planet". Using "galactic star formation histories, metallicity distributions, and the likelihood of stars hosting Earth-like planets in their habitable zones, under specific assumptions which we describe as the Astrobiological Copernican Weak and Strong conditions", the researchers conclude that the nearest incidence of intelligent civilizations far surpasses "our ability to detect it for the foreseeable future, and making interstellar communication impossible".[16]

Unsurprisingly, a comprehensive search of a patch of the Southern sky by astronomers working at the International Centre for Radio Astronomy Research (ICRAR) in Australia, equally yielded "not even a hint of alien technology" at low radio frequencies, according to a report on the *ScienceAlert* website in September 2020. One of the researchers concluded: "Since we can't really assume how possible alien civilisations might utilise technology, we need to search in many different ways."[17]

According to SETI Institute's Jill Tarter, whose work was portrayed by Jodie Foster's character in the film *Contact* (1997), a common misconception is that the search for extraterrestrial intelligence (SETI) and UFOs are in any way related. She says: "SETI uses the tools of the astronomer to attempt to find evidence of somebody else's technology coming from a great distance. If we ever claim detection of a signal, we will provide evidence and data that can be independently confirmed. UFOs – none of the above."[18]

Notwithstanding Dr Tarter's dismissal, there has been notable movement on the science front. Although there is no direct physical evidence for the extraterrestrial hypothesis for the UFO phenomenon, professors Wendt and Duvall maintain, "there is considerable *in*direct physical evidence for it, in the form of UFO anomalies that lack apparent conventional explanations – and for which ETs are therefore one possibility. (…) Such anomalies cannot be dismissed simply because they are only indirect evidence for ETs, since science relies heavily on such evidence, as in the recent discovery of over 300 extra-solar planets (and counting). For if UFO anomalies are not potentially ETs, what else are they?"[19]

In fact, in a paper from 2018 Dr Silvano Colombano at the NASA Ames Research Center, a SETI Institute partner, says: "It seems to me that SETI has ignored (at least officially) the potential relevance of UFO phenomena for three reasons: (1) The assumption of extremely low likelihood of interstellar travel, (2) The very high likelihood of hoaxes, mistaken perceptions or even psychotic events in UFO phenomena, and (3) The general avoidance of the subject by the scientific community." In response, Dr Colombano proposes a more "aggressive" approach to the search for extraterrestrial life. Writing about advanced civilizations he says that, given that our current scientific methodologies have developed over only the last 500 years or so, "we might have a real problem in predicting technological evolution even for the next thousand years, let alone 6 million times that amount!"

Nevertheless, he suggests, we should consider the UFO phenomenon worthy of study in the context of a system with very low signal-to-noise ratio, "with the possibility of challenging some of our assumptions and pointing to new possibilities for communication and discovery".[20] Our challenge, then, is to establish a system with the highest possible signal-to-noise ratio, by collecting evidence from as wide a range of relevant disciplines that provide mutual corroboration.

The answer to the question "what are they?" – and subsidiary questions like "how fast did it go?", "how large was it?", "what is it made of?" and such more – are considered essential by 'serious' researchers to come to an acceptable understanding of, or explanation for, the UFO phenomenon. Faced with the obviously out-of-this-world capabilities of many sighted objects these are certainly valid questions whose answers would fascinate anyone with an interest in advanced technologies. Yet, enquiring into the physical aspects of a phenomenon that has proven so elusive, despite its persistence in the face of our advancing scientific methods, is itself based on assumptions that are challenged by a growing number of scientists.

In the Galileo Commission Report (2019) Dr Harald Walach provides an incisive analysis in this respect. He points out that "the basic assumption of modern science is that matter is the most fundamental entity in the Universe"[21], and everything we see and experience is a product of matter. But the deeper physicists are able to see into the nature of matter, Dr Walach shows, the less material their findings get, and the more confirmation they find of what previously were esoteric notions of reality. Matter turns out to be not as fundamental as the predominant scientific view would have us believe. When one claims, he says, that the universe originated when "matter emerged spontaneously out of an incredibly dense energy, which itself emerged out of … immaterial informational blueprints" the very notion of 'blueprints' implies a deeper structure and level of reality from which the 'blueprints' originate.

He then proceeds to present evidence for his argument that consciousness does not emerge from a particularly fortunate amalgam of material particles, and is not dependent on brain activity, but plays its own causal role in the manifestation of what we call reality.[22] Dr Walach provides various examples, like for instance what is known as Schrödinger's entanglement theory, which says that if you act on a particle here, the action has an instant effect on a particle far away. Einstein referred to this effect as 'spooky action at a distance'. Physicists have been able to test this theory, which is now accepted as reality.[23]

In fact, according to professor of Theoretical Physics at Cambridge University David Tong in a talk in February 2017, "the very best theories we have of physics don't rely on particles at all. (…) The fundamental building blocks of nature are fluid-like substances which are spread throughout the entire universe and ripple in strange and interesting ways. That's the fundamental reality in which we live."[24]

Let us acknowledge here that, long predating the scientific discovery that there is no such thing as an ultimate building block of matter, it was Helena Petrovna

Blavatsky who wrote in 1888: "It is on the doctrine of the illusive nature of matter, and the infinite divisibility of the atom, that the whole science of Occultism is built."[25] With her work for the Theosophical Society, of which she was a co-founder, Mme Blavatsky laid the foundation for the reinstatement of the Ageless Wisdom teaching in modern times.

So the particles that physics considers the building blocks of life aren't really particles at all. They are waves in a sea of possibilities that are tied up in small bundles of energy. These discrete bundles, or 'quanta', are currently the smallest that our science is able to detect; how they behave, will behave or could behave belongs to the field of quantum mechanics, quantum physics, and quantum theory. In his recent book *The Intelligence of the Cosmos* professor Ervin Laszlo summarizes the latest findings in this regard: "The universe, as we now know, is not a domain of matter moving in passive space and indifferently flowing time; it is a sea of coherent vibrations."[26]

Another significant note to be made here is that in 1932 a teacher of metaphysics wrote in strikingly similar terms: "I am in the invisible ocean of vibrations or consciousness." And later: "We must remember that consciousness is the sea of life within which all forms are living, regardless of what they may be."[27] This teacher was George Adamski, the Polish American who was the first to speak out about his contact with a visitor from space in November 1952 (see page 1) and who became the subject of the most enduring character assassination attempt as a result.[28]

According to Dr Walach, "The viability of materialism as a background philosophy of science is contingent on the success of a materialist theory of mind and consciousness."[29] And consciousness, as his report shows, "is categorically different from all material systems we know"[30], which means that, while lacking fundamental particles to experiment with, the deeper 'structure' and 'reality' – the web of consciousness in which we are innately connected with everything – for now can only be accessed by humans through experience. In other words, the materialist scientific view is too limited to explain consciousness, and therefore reality.

Even in 1956 US contactee Howard Menger was told by a visitor from space whom he referred to as a "profound space teacher" that, although science is necessary for making progress in all aspects of life, it is now "actually limiting itself, and the progress of your populace, by sanctioning that which it is able to prove only through objective reality, rather than subjective reality as such, related to truth. A scientist observes visually in the course of a particular experiment an occurrence which did take place before his eyes, but, because he did not know how or why it happened, he rejects it, and it is not 'scientific' fact until he finds the answer within the scope of accepted scientific theory."[31]

What if consciousness is not some peculiar brain activity, but is instead a quality inherent in all matter, asks philosopher Philip Goff in an interview with *Scientific American*: "Yes, physical science has been incredibly successful. But it's been successful precisely because it was designed to exclude consciousness. If Galileo were to time travel to the present day and hear about this problem of explaining consciousness in the terms of physical science, he'd say, 'Of course, you can't do that. I designed physical science to deal with *quantities,* not *qualities.*' "[32]

Even Isaac Newton, whose *Mathematical Principles of Natural Philosophy* is often cited as the source of the mechanistic view of the solar system, in its final paragraph referred to "something concerning a certain most subtle spirit which pervades and lies hid in all gross bodies", which causes particles to be attracted or repelled to make up all the various discrete forms. "But these are things that cannot be explained in few words, nor are we furnished with that sufficiency of experiments which is required to an accurate determination and demonstration of the laws by which this electric and elastic spirit operates."[33]

In other words, we need alternative ways of approaching and understanding reality than merely reducing it to its material components for analysis. And this is precisely where the current approach summarized by Dr Roush in his survey *Extraterrestrials* falls short – it leaves out an aspect of reality that the best minds in quantum science say is more fundamental than our material reality.

Dr Colombano already suggested we look in more directions in our search for extraterrestrial life, by engaging physicists in "speculative physics", technologists in futuristic exploration of how technology might evolve, and sociologists in speculation about what kinds of societies we might expect from such developments.[34]

Just looking at the latest findings from science itself, life seems rather more abundant than it is generally given credit for. For instance, based on recent research astrobiologists have shown that "if the origin of life can occur rather easily, we should live in a cosmic zoo, as the innovations necessary to lead to complex life will occur with high probability given sufficient time and habitat. On the other hand, if the origin of life is rare, then we might live in a rather empty universe."[35]

Separate findings indicate that life is more likely the rule than the exception in the universe. 3.7 Billion-year-old fossils of microorganisms indicate that life originated more than four billion years ago in the Hadean eon, before the earliest rocks were formed, while Earth itself only formed 0.9 billion years prior. Dr Abigail Allwood, of NASA's Jet Propulsion Laboratory, concluded: "Give life half an opportunity and it'll run with it. Our understanding of the nature of life in the Universe is shaped by how long it took for Earth to establish the planetary conditions for life. If life could find a foothold here, and leave such an imprint that vestiges exist even though only a minuscule sliver of metamorphic rock is all that remains from that time, then life is not a fussy, reluctant and unlikely thing."[36]

Dr Allwood's conclusion recently found support in a research paper by David Kipping, assistant professor in Columbia University's Department of Astronomy, who used a statistical technique to shed light on how complex extraterrestrial life might evolve on other worlds. While he says his analysis cannot provide certainties or guarantees, "the case for a universe teeming with life emerges as the favored bet."[37]

As the possibility of extraterrestrial visitors is slowly – ever so slowly – gaining acceptance, scientific evidence supporting the notion that life is not confined to planet Earth seems to be mounting more noticeably. But, as physicist Enrico Fermi's Paradox states: If probability tells us there should be millions of civilizations in the Universe, why haven't we found any of them?

On 17 June 2012, when the International Space Station was passing over the People's Republic of China, a teleconference was set up between China's CCTV (Central China Television) channel and the crew of the Russian segment of the ISS, Gennady Padalka, Sergey Revin and Oleg Kononenko. The cosmonauts communicated with Chinese television viewers and answered numerous questions from the public. Particularly interesting was the discussion on the possibility of meeting extraterrestrial beings, when cosmonaut Padalka replied that "in the Universe humanity cannot be alone, sooner or later we will meet brothers in mind". He also reminded CCTV viewers of the existence of a detailed United Nations instruction in case of first contact.[38]

Writing about the high expectations of finding life elsewhere in the early days of the space age, George Adamski said: "While we may not be allowed to go into space with a warlike or hostile intent, we will be helped out far enough to see for ourselves that life as we know it exists elsewhere in the system. Let us hope the public is allowed to hear what the first men into outer space report back to earth."[39]

Although downplayed or derided, the stories of astronauts who have gone on the record about having witnessed unearthly craft are well-known and include the names of Apollo 14 astronaut Edgar Mitchell[40], Gemini 5 astronaut Gordon Cooper[41], Mercury 7 astronaut Deke Slayton[42], and Brian O'Leary, who was one of eleven candidates for a possible mission to Mars in 1967, and went on to become a respected professor of Physics at Princeton University. Shortly before his passing in 2012 he said in an interview: "There is abundant evidence that we are being contacted. Civilizations have been monitoring us for a very long time and that their appearance is bizarre from any type of traditional materialistic western point of view."[43] Strikingly, a very similar view was shared by former deputy director of the Bulgarian Space Research Institute Lachezar Filipov in 2009, who stated: "Aliens are currently all around us, and are watching us all the time. They are not hostile towards us, rather, they want to help us but we have not grown enough to establish direct contact with them."[44] In keeping with the 'scientific' taboo on the subject, he was subsequently stripped of his academic positions.

In January 2020 the chorus of pilots and astronauts – professions not known for their susceptibility to figments of the imagination – who have spoken out about their experience with or conviction of the reality of extraterrestrial visitors added another member in Dr Helen Sharman, the first British astronaut in space. In an interview with *The Observer* Dr Sharman said: "Aliens exist, there's no two ways about it. There are so many billions of stars out there in the universe that there must be all sorts of forms of life. Will they be like you and me, made up of carbon and nitrogen? Maybe not. It's possible they're right here right now and we simply can't see them."[45]

Regarding the question "If we're not alone, where are they?" it is interesting that Dr Sharman raises the possibility that "we simply can't see them". As I pointed out elsewhere, based on its calculations of the mass of the Universe astrophysics itself admits that it doesn't know what more than 90 per cent of the cosmos consists of.[46] For decades it has hypothesized 'dark matter' and 'dark energy' to explain this 'missing' mass of the Universe. Coincidentally, dark matter was first theorized by Swiss astronomer Fritz Zwicky who worked at the Palomar Observatory and, though sceptical, visited George Adamski at the Palomar Gardens Cafe three times, but publicly ridiculed him.[47]

So what would we find if a broader approach to our search for proof of extraterrestrial life or visitors included sources that have thus far been dismissed by materialist science? As the findings of quantum scientists are finally catching up with the esoteric wisdom teachings, the latter may help fill in the astrophysical knowledge gap.

For example, the teachings tell us that above the solid, liquid and gaseous physical levels of matter that fall within our range of vision, there are four further planes of matter where the subatomic particles consist of light, as they vibrate at higher frequencies than those on the lowest three planes.[48] And already, mainstream astronomy is edging closer to this esoteric distinction between planes of dense and subtle matter, with findings being reported which "suggest that dark matter is another kind of sub-atomic particle, possibly forming a parallel universe of 'supersymmetry' filled with supersymmetrical matter that behaves like an invisible mirror-image of ordinary matter."[49] In fact, if we are ready to look beyond the strictly materialist scientific practices, and look in more directions as advocated by Dr Colombano, tentative evidence for the existence of higher planes of (subtle) matter may perhaps already be found in the pioneering work of Wilhelm Reich MD, who discovered what he called 'orgone radiation'; Semyon Kirlian, who developed the technique to photograph human auras; and Rupert Sheldrake PhD, who proposed and experimented with the hypothesis of formative causation, which suggests a kind of memory bank from where Nature retrieves its blueprints for the physical forms of species.

If life does not depend exclusively on carbon-based dense-physical forms to manifest, these findings would make the claims of the 1950s flying saucer contactees that they were contacted by visitors from space seem not so outlandish after all. Indeed, one of the teachers of the wisdom philosophy wrote in the early 1880s: "Our planet (like all those we see) is adapted to the peculiar state of its human stock, that state which enables us to see with our naked eye the sidereal bodies which are co-essential with our terrene plane and substance, just as their respective inhabitants, the Jovians, Martians, and others, can perceive our little world; because our planes of consciousness, differing as they do in degree, but being the same in kind, are on the same layer of differentiated matter…"[50]

Another wisdom teacher, writing in 1925, specified that the etheric planes of matter "are but gradations of physical plane matter of a rarer and more refined kind, but physical nevertheless" and names them, from the highest to the lowest plane, as (1) atomic matter, (2) sub-atomic matter, (3) super-etheric matter, and (4) simply etheric matter. This fourth plane, he says, is the only one that scientists have recognised and investigated.[51]

Therefore, considering that the esoteric wisdom tradition states that life on other planets in our system doesn't precipitate down from these subtle planes of matter onto the dense physical planes as it does at this stage in the evolution of our planet, it wouldn't require a leap of faith to see that visitors could also hail from within our own solar system. Or that, when extraterrestrial craft or visitors allow themselves to be seen by us, they temporarily lower the rate of vibration of their composing elements from the subtle physical into the dense physical, so as to fall within our range of vision. And thus far, this seems to be the only viable response to the Fermi Paradox.

Science, at least the more enlightened variety that is not stuck in the rut of dense-physical reductionism, increasingly indicates that to grasp the reality or the significance of the UFO phenomenon our focus should not be on the technology, but on that of which the technology is a manifestation. So it is only fitting that over the last decade or so another strand of 'serious' research has gained much in stature and visibility, which holds that 'consciousness' is a distinct factor to be included in UFO and extraterrestrial intelligence research.

For her book *American Cosmic* Dr Diana Pasulka, professor of religious studies at the University of North Carolina, interviewed well-known and prominent scientists, professionals, and hi-tech entrepreneurs to show that the belief in UFOs is not limited to a fringe audience. She examines the mechanisms that lie behind people's interpretation of unexplainable experiences, based on her contention that "widespread belief in aliens is due to a number of factors including their ubiquity in modern media like The X-Files…"

Equating people's experiences with a new belief, she writes: "The infrastructure of technology has spawned new forms of religion and religiosity, and belief in UFOs has emerged as one such new form of religious belief."[52] Given the examples and people that she quotes, Dr Pasulka's understanding of consciousness still seems to have a material foundation, with consciousness emerging from or dependent on technology, for instance when she cites UFO 'authority' Jacques Vallee: "Consciousness could be defined as the process by which informational associations are retrieved and traversed."[53]

Ignoring the question what, then, would be the *source* of the 'information' or the *agent* of the 'association', Dr Pasulka unwittingly falls into the old materialist trap, which even a more comprehensive definition of 'religion' might have helped her avoid. Instead of merely distinguishing between functional aspects of religion such as rituals and what she calls "the sacred element" in her book[54], if we abstract the essence from the particulars of the various world religions we find they all share the same three tenets: the cyclical coming or return of a Teacher (Messiah, Second Coming, fifth Buddha, tenth incarnation of Vishnu, twelfth Mahdi) who brings a new revelation about the source (a personal God) and evolution (return to God) of consciousness, which needs to be given expression in the establishment of right human relations (the Golden Rule).[55]

In essence, therefore, religion, stripped of centuries of man-made dogma, is not a belief, but a technique to reconnect (or, in original Latin, *religare*) with the Source of life and consciousness. It can be no accident that *yogam*, the Sanskrit root of the word yoga, also means to reconnect. In fact, Margit Mustapa, a little-known contactee from Finland who emigrated to the US in 1947, was told by her contact from space: "Religion means for us a unified process of radioactivity, which is telepathy between man and God."[56]

While Dr Pasulka does refer to those who claim to have experienced some form of contact with non-humans as 'contactees', nowhere does she acknowledge the fact that the notion of extraterrestrials visiting Earth was initially normalized by the abundance of accounts and experiences of contactees around the world in the 1950s.

A major effort with a clear focus on the consciousness aspect of contact was conducted by the Foundation for Research into Extraterrestrial and Extraordinary Experiences (FREE). Co-founded by the late Apollo 14 astronaut Dr Edgar Mitchell, FREE conducted three surveys involving thousands of 'contact experiencers' from over 100 countries. The findings from their qualitative and quantitative research were published in the 2018 report *Beyond UFOs: The Science of Consciousness and Contact with Non Human Intelligence*. And although the 780-page FREE report does look into a few of the original contact cases, its focus is almost exclusively on the accounts of later 'experiencers'.

However, now that consciousness research advances and expands our understanding of reality, George Adamski's accounts and those of the other 1950s contactees can no longer be dismissed off-hand and should not be ignored. These original accounts of contact became public before the contact narrative was contaminated by tales of abduction, mutilation, and other fear-inducing atrocities in order to confuse and frighten the public who showed great interest in the original message of peace and international cooperation in the midst of the Cold War. Therefore, qualitative research and a critical synthesis of the correspondences among these sources provide an indispensable touchstone to scrutinize later claims of contact for possible contamination. This is all the more important given the long-standing confusion in the field since it was infected with misinformation. As I have shown elsewhere, in addition to contact with extraterrestrial visitors, without proper discernment claims of contact could well relate to other phenomena, like astral entities, mediumistic channellings, out-of-body experiences, drug-induced staged experiences, figments of people's overactive imagination, et cetera.[57]

Yet, ignoring the tell-tale characteristics of original contact experiences, FREE's research lumps all these together under the label of contact with "non-human intelligence". But without the necessary discernment they would more aptly be called experiences of non-corporeal or "non-local" consciousness. According to Harald Walach non-locality may be described as "connections between domains of space and time, or between conscious minds across space and time, that do not rely on known signals or occur without signal transfer"[58], e.g. near-death experiences, out-of-body experiences, telepathy, channelling, et cetera, which could involve human, astral, super-human, or extraterrestrial consciousness.

When Dr Walach writes about post-materialist science that "Consciousness has to be taken seriously in its own right and not only as a potential emergent of a complex neuronal system"[59], his evidence-based postulate confirms what Madame Blavatsky wrote almost 150 years prior: "Theosophists (…) are the first to recognize the intrinsic value of science. But when its high priests resolve consciousness into a secretion from the grey matter of the brain, and everything else in Nature into a mode of motion, we protest against the [scientific] doctrine as being unphilosophical, self-contradictory, and simply absurd, from a scientific point of view…"[60]

Physicist David Bohm (1917-1992) was one of the first scientists to predict the non-locality of consciousness or, simply put, the interconnectedness of the Universe. But, in striking coincidence with the fate of the message that the contactees were asked to share with the world, Bohm's progressive ideas were considered a threat to the scientific establishment who at that time were focused only on the atomic bomb.[61]

In the conclusion to the FREE report, the authors write: "A discipline of human endeavor based on research of such personal CEs [contact experiences], which have been largely ignored by the scientific, psychiatric, and parapsychological communities, and by ufology and abduction researchers, may present a new paradigm of human transformation and transcendence which may eventually evolve towards a greater understanding of ourselves in the universe, consciousness, and possibly even reality itself."[62]

Being the first to posit the notion of the supremacy of consciousness over matter, and the evolution of consciousness as the drive behind the manifestation of life, however, Blavatsky's work presents precisely the "new paradigm of human transformation and transcendence" that Dr Klimo and his fellow authors are anticipating.

And although the FREE researchers rightly note that most researchers have ignored contact experiences, they in turn make the same mistake by ignoring the wisdom teachings, despite Jon Klimo, one of the authors, referencing the work of H.P. Blavatsky, Alice Bailey, and Benjamin Creme in his earlier book *Channeling. Investigations on Receiving Information from Paranormal Sources* (1987). Even if the method by which they received their information isn't understood, the fact that the basic tenets of their work are finding more and more confirmation in systems science and quantum research sets it apart from mediumistic messages. However, like most academicians, here too Dr Klimo fails to distinguish between mental telepathy – the deliberate, conscious communication between two minds on the mental plane – and the random channeling of astral thoughtforms, thereby foregoing a solid criterion for a useful categorization of the different types of experience in FREE's research or for filtering out irrelevant data.

Another reason for leaving out the esoteric teachings on consciousness may be that Madame Blavatsky was just as strongly vilified and denounced in her day as were George Adamski and the other contactees of the initial 'flying saucer' era. But just as proper scrutiny of the allegations against him clears Adamski of fraud and deception[63], so the Society for Psychical Research who labelled Blavatsky a fraud in an 1885 report, retracted their condemnation a full century later.[64]

"Intelligent civilizations would be based on carbon life"

24

and "We have not been, and are not being... visited" are just some of the most widely held assumptions that Dr Colombano challenges in his paper. He says: "In the very large amount of 'noise' in UFO reporting there may be 'signals', however small, that indicate some phenomena that cannot be explained or denied."[65] While he does not explicitly mention quantum theory or consciousness studies as directions to include in the search for intelligent extraterrestrial life, the latest research presented here provides ample reason why these should be taken into account.

During a talk in Belgium in May 1963, George Adamski already implied that a broader approach to UFO research was essential, when he said: "All they are doing is reporting sightings. You have a sighting, you report it, they put it out, that's all. That is not the purpose. We got to do more than that. And so they are not getting anywhere, except confusion; but the truth really is that the officials know a lot more."[66] And since governments have kept their knowledge and contacts secret, we should begin by including the information from the contactees to allow a broader perception to be formed, rather than having it obscured by the disinformation that was meant to discredit them.

Filtering out much of the 'noise' by looking for corroborations across accounts and from other disciplines, including the wisdom teaching, my previous books show that much may be learned about the reality of extraterrestrial visitors from these accounts of the original contactees.

At the forefront of science, professor Laszlo confirms that nature, the world, or reality is not a construction of discrete bits and pieces of matter that have been put together as some mechanism, which we could take apart and put back together in a different way: "If you look at it in that way you'll end up by destroying that unity which is their nature. (…) There is this factor in nature, which means that every living, surviving species is somehow built into its environment. (…) only humans, in the scope of the last 100-150 years or so, have managed to be outside of this system, not aligning ourselves in it, or with it, and acting only in view of our own immediate perceived interest and we are compensating for that by technology. But in the process, we are destroying our environment, the natural balances in the environment, and we're also overloading the system in terms of number and energy and resources, resource claims, and so on."[67]

Four years later, in June 2020, professor Laszlo's warning is echoed in a call for urgent action coordinated with the UN's World Health Organisation (WHO) and the UN Convention on Biological Diversity, in the World Wildlife Fund (WWF) report *Covid-19: Urgent Call to Protect People and Nature*: "COVID-19 is a devastating wake-up call that humanity's broken relationship with nature affects not only the wildlife and natural ecosystems whose habitats are being destroyed, but also threatens human health. By continuing to damage natural habitats, humans risk incurring the terrible costs of new zoonotic diseases, as well as increased exposure to other threats such as climate change."[68]

Like many, Gus Speth, an environmental lawyer and US government advisor under presidents Carter and Clinton, used to think that top environmental problems were biodiversity loss, ecosystem collapse and climate change: "I was wrong. The top environmental problems are selfishness, greed and apathy, and to deal with these we need a cultural and spiritual transformation. And we scientists don't know how to do that."[69]

In the announcement for an online lecture for the British Royal Institution on 16 June 2020 astrophysicist Martin Rees unintentionally provides a case in point. He first acknowledges that humanity "has reached a critical moment. Our world is unsettled and rapidly changing, and we face existential risks over the next century" as "the future of humanity is bound to the future of science". But apparently oblivious to the advances in systems science and quantum research, his talk explored how this future "hinges on how successfully we harness technological advances to address our challenges"[70], rather than how we base our technological advances on the awareness that life is one and interconnected.

The facts about the impending threats to planet Earth, in whose manifestation humanity is the first stage of individual self-consciousness, more than anything else, should make it clear that the current materialist-based approach to reality has set us on the path of self-destruction, and we urgently need to take a broader view to understand how our world, our lives, our consciousness and our destiny are intricately connected, and therefore depend on correctly relating to each other and our surroundings.

Modern transport, information and communications technology, and scientific discoveries have changed our experience of the physical world to that of an interconnected global community. Our political, economic, financial, and social structures, however, still reflect a worldview based on division and competition for territory, power and wealth. This chasm represents the crisis of consciousness that humanity is currently facing, which is reflected in unprecedented levels of inequality and social injustice, globally and domestically, with all the associated dangers of self-destruction. For this reason, George Adamski was told – as were many other contactees in similar terms: "Now that your scientific knowledge has so far outstripped your social and human progress, the gap between *must* be filled with urgent haste."[71]

Historian Arnold Toynbee called this the morality gap: "Technology gives us material power – the greater our material power, the greater our need for the spiritual insight and virtue to use power for good and not for evil. The 'morality gap' means that (…) we have never been adequate spiritually for handling our material power. Today it is greater than ever."[72]

Moral philosopher Toby Ord thinks humanity "is in its adolescence, and like a teenager that has the physical strength of an adult but lacks foresight and patience, we are a danger to ourselves until we mature." In his interview with *The Observer* about the Covid-19 crisis, he recommends that "we slow the pace of technological development so as to allow our understanding of its implications to catch up and to build a more advanced moral appreciation of our plight."[73]

All things considered, therefore, the proverbial nuts and bolts and how they are put together are only the technological manifestation of a particular level of understanding of the physical world – a particular level of consciousness at any given time. The discovery of fire, agriculture, steam engine, rocket science, et cetera are all expressions of humanity's awareness, understanding, and experience of the world we lived in at that time. So, when we limit our search to the 'nuts and bolts', the material or technological aspect of the UFO phenomenon or extraterrestrial life, we approach the subject from the wrong end of the stick, from the effects instead of the cause end of the stick.

Wilbert Smith was a Canadian engineer and contactee, who had earned an M.Sc. degree in Electrical Engineering and held several patents, but also had a deep interest in philosophy and religion. In *The New Science*, which he wrote in the final years of his life, he ponders the relationship between reality and awareness (consciousness): "...we do know, or we think that we know, that Reality does exist and we are aware of it. But we do not know how far beyond us the Awareness extends, and we must either postpone this determination pending a better understanding or accept the statements by other entities who are presumably more advanced than we are that Awareness is universal and extends throughout all Reality..." And, "as our understanding increases, we are able to devise ways and means for extending our senses both in range and scope, which in turn leads to better understanding. But we must always remember in thus extending our senses to *distinguish between the language of the observation and its translation into the language of our senses, lest we miss the phenomenon while inspecting its effect.*"[74] [emphasis added]

In other words, we need to see that understanding life and the world is about *our* understanding of life and the world. This consciousness cannot be understood by physical experiments and, as Dr Walach suggests, "the basic intuition of reality itself might actually transcend classical, binary logic"[75], for which he cites various examples of scientific discoveries that were arrived at through such "radical introspection". Einstein said of this process: "There comes a time when the mind takes a higher plane of knowledge but can never prove how it got there. All great discoveries have involved such a leap. The important thing is not to stop questioning."[76] George Adamski seems to refer to a similar process when he writes: "Some of us attain this state by studying sincerely and truthfully, while to others it is a natural thing to do without any study. This latter seems to be the case with me. Things just simply come to me and I am able to stay with them until the full revelation has been completed. (…) I still do not deserve any credit since all this knowledge belongs to all the universe and not to anyone in particular. The ONLY credit that ANY form may deserve is that it becomes a willing form through which such revelations may come, whereby others may advance."[77]

Hence, in order to increase our understanding of life, the world we live in, and the universe, we need to acknowledge that life is about the *evolution of consciousness itself*, the growing awareness of the underlying oneness of life and the interconnectedness of its composing elements, so that our advancing scientific insights and technological knowledge may serve the common good of all members of humanity and the planet alike.

Quantum theory, consciousness research, and systems science now reveal Mme Blavatsky and George Adamski as pioneers in their respective fields and we cannot ignore their trailblazing efforts without perpetuating the prevalent confusion and misunderstanding about the nature and origin of UFOs and the reason for the extraterrestrial visitations.

References

1 Helene Cooper, Ralph Blumenthal, and Leslie Keane, 'Glowing Aura's and 'Black Money': The Pentagon's Mysterious U.F.O. Program', *New York Times*, 16 December 2017. See: <www.nytimes.com/2017/12/16/us/politics/pentagon-program-ufo-harry-reid.html>

2 Alexander Wendt and Raymond Duvall, 'Sovereignty and the UFO', in *Political Theory*, Vol.36, No.4, August 2008, p.610. See: <journals.sagepub.com/doi/pdf/10.1177/0090591708317902>

3 J. Allen Hynek (1972), *The UFO Experience: A Scientific Inquiry*, p.180

4 Ibidem, p.103

5 Cooper, Blumenthal, Keane (2017), op cit

6 Cooper, Blumenthal, Keane, ' "Wow, What Is That?" Navy Pilots Report Unexplained Flying Object', *New York Times*, 26 May 2019. See: <www.nytimes.com/2019/05/26/us/politics/ufo-sightings-navy-pilots.html>

7 Joe Montaldo (2020), 'UFO Undercover Guests UFO lobbyist Stephen Bassett'. Live stream 8 July 2020. See: <youtu.be/zoYN3T9wnQ4?t=5288>

8 M.J. Banias and Tim McMillan, 'The Navy Says the UFOs in Tom DeLonge's Videos Are 'Unidentified Aerial Phenomena', *Vice*, 17 September 2019. See: <www.vice.com/en_ca/article/8xw83b/the-navy-says-the-ufos-in-tom-delonges-videos-are-unidentified-aerial-phenomena>

9 Daniel Strauss, 'Pentagon releases three UFO videos taken by US navy pilots', *The Guardian*, 28 April 2020. See: <www.theguardian.com/world/2020/apr/27/pentagon-releases-three-ufo-videos-taken-by-us-navy-pilots>

10 Blumenthal, Keane, 'No Longer in Shadows, Pentagon's U.F.O. Unit Will Make Some Findings Public'. *New York Times*, 23 July 2020. See: <www.nytimes.com/2020/07/23/us/politics/pentagon-ufo-harry-reid-navy.html>

11 Ryan Browne, 'Pentagon to launch task force to investigate UFO sightings'. CNN, 14 August 2020. See: <edition.cnn.com/2020/08/13/politics/pentagon-ufo-task-force/index.html>

12 Wendt, Duvall (2008), op cit, p.611

13 Lee Speigel, 'WikiLeaks Documents Reveal United Nations Interest in UFOs', *Huffington Post*, 28 October 2016. See: <www.huffpost.com/entry/wikileaks-ufos-united-nations_n_5813aa17e4b0390e69d0322e>

14 Red Panda Koala (Josh Anderson), 'United Nations UFO Disclosure', 15 January 2020. See: <www.youtube.com/watch?v=MWKysekGJU8>

15 Wade Roush (2020), *Extraterrestrials*, pp.110-11.

16 Tom Westby and Christopher J. Conselice, 'The Astrobiological Copernican Weak and Strong Limits for Intelligent Life'. *The Astrophysical Journal*, 15 June 2020. See <iopscience.iop.org/article/10.3847/1538-4357/ab8225>

17 Michelle Starr, 'Across Ten Million Stars Not a Single Whisper of Alien Technology'. *ScienceAlert*, 8 September 2020. See: <www.sciencealert.com/across-10-million-stars-not-a-single-whisper-of-alien-technology-has-been-detected>

18 Leonard Tramiel, 'Life As We Know It: An Interview With Jill Tarter'. *Skeptical Inquirer*, 26 August 2016. See: <skepticalinquirer.org/exclusive/life-as-we-know-it-an-interview-with-jill-tarter/>

19 Wendt, Duvall (2008), op cit, pp.614-15

20 Silvano P. Colombano Ph.D., 'New Assumptions to Guide SETI Research', 15 March 2018. See: <ntrs.nasa.gov/archive/nasa/casi.ntrs.nasa.gov/20180001925.pdf>

21 Harald Walach (2019), *Beyond a Materialist Worldview. Toward an Expanded Science*, p.29. See: <www.galileocommission.org/wp-content/uploads/2020/02/Science-Beyond-A-Materialist-World-View-Digital-18.02.20.pdf>

22 Ibidem, p.64

23 Ibid, pp.65-66

24 David Tong, 'Quantum Fields: The Real Building Blocks of the Universe'. The Royal Institution lecture, 15 February 2017. See: <www.youtube.com/watch?v=zNVQfWC_evg>

25 H.P. Blavatsky (1888), *The Secret Doctrine*, Vol.I, p.520 (6th Adyar ed. Vol.2, p.244)

26 Ervin Laszlo (2017), *The Intelligence of the Cosmos*, p.21

27 George Adamski (1932), *The Invisible Ocean*, pp.10-11; and (1964), *Science of Life* study course, Lesson Eleven, 'Exploration of Cosmic Space'

28 Desmond Leslie and George Adamski (1953), *Flying Saucers Have Landed*, p.185ff

29 Walach (2019), op cit, p.44

30 Ibid, p.49

31 Howard Menger (1959), *From Outer Space to You*, p.176

32 Gareth Cook, 'Does Consciousness Pervade The Universe? Interview with Philip Goff'. *Scientific American*, 14 January 2020. See: <www.scientificamerican.com/article/does-consciousness-pervade-the-universe/>

33 Isaac Newton (1846), *The Mathematical Principles of Natural Philosophy*, Book III: General Scholium (English translation by Andrew Motte), p.507

34 Colombano (2018), op cit

35 William Bains and Dirk Schulze-Makuch, 'The (Near) Inevitability of the Evolution of Complex, Macroscopic Life'. MDPI, 30 June 2016. See: <www.mdpi.com/2075-1729/6/3/25/htm>

36 Ian Johnston, 'World's oldest fossils found in discovery with 'staggering' implications for search for alien life'. *The Independent*, 31 August 2016. See: <www.independent.co.uk/news/science/oldest-fossils-world-alien-life-earth-mars-greenland-a7218191.html>

37 David Kipping, 'An objective Bayesian analysis of life's early start and our late arrival'. *Proceeding of the National Academy of Sciences*, 18 May 2020. See: <www.pnas.org/content/early/2020/05/12/1921655117>

38 'Russian Cosmonaut: Sooner or later, humanity will meet "brothers in mind" ', 19 June 2012. See: <ufology-news.com/novosti/po-mneniyu-rossijskogo-kosmonavta-chelovechestvo-rano-ili-pozdno-vstretit-bratev-po-razumu.html>

39 C.A. Honey (ed.), 'Questions and answers', *Cosmic Science* newsletter Vol.1 No.1, January 1962, p.5

40 Interview with Edgar Mitchell in Nick Margerrison, *The Night Before*, Kerrang! Radio, 23 July 2008. See: <www.youtube.com/watch?v=RhNdxdveK7c>

41 Interview with Gordon Cooper in Mack Anderson and Bradley Anderson, *Paranormal Borderline. Breaking the Silence*, 7 May 1996. See: <youtu.be/VfsoOVvV5Ns>

42 Interview with Deke Slayton in Robert Guenette (1985), *America Undercover. UFOs: What's Going On?* See: <youtu.be/4E_SFwrh8wo>

43 Interview with Brian O'Leary in Steve Gagné and Kimberly Carter Gamble, *Thrive*, 2011. See: <youtu.be/yO0T05kQkbs>

44 'Aliens "already exist on earth", Bulgarian scientists claim', *The Telegraph*, 26 November 2009. See <www.telegraph.co.uk/news/worldnews/europe/bulgaria/6650677/Aliens-already-exist-on-earth-Bulgarian-scientists-claim.html>

45 Michael Segalov, 'There's no greater beauty than seeing the Earth from above'. Interview with Helen Sharman, *The Observer*, 5 January 2020. See: <www.theguardian.com/lifeandstyle/2020/jan/05/astronaut-helen-sharman-this-much-i-know>

46 Ian Sample, 'Scientists hunt mysterious "dark force" to explain hidden realms of the cosmos'. *The Guardian*, 3 September 2018. See: <www.theguardian.com/science/2018/sep/03/scientists-hunt-for-dark-force-todiscover-what-the-universe-is-made-of>

47 Don Lago, 'Messages from Space'. *Michigan Quarterly Review*, Vol.54, No.1, Winter 2015. See: <hdl.handle.net/2027/spo.act2080.0054.108>

48 Alice A. Bailey (1950), *Telepathy and the Etheric Vehicle*, pp.142-43

49 Steve Connor, 'The galaxy collisions that shed light on unseen parallel Universe'. *The Independent*, 26 March 2015. See: <www.independent.co.uk/news/science/the-galaxy-collisions-that-shed-light-on-unseen-parallel-universe-10137164.html>

50 Blavatsky (1888), op cit, Vol.I, p.166 (6th Adyar ed. Vol.1, p.220)

51 Bailey (1925), *A Treatise on Cosmic Fire*, p.910

52 D.W. Pasulka (2019), *American Cosmic – UFOs, Religions, Technology*, p.15

53 Jacques Vallee, as quoted in Pasulka (2019), op cit, p.176

54 Pasulka (2019), op cit, p.242

55 Gerard Aartsen, 'A global crisis in consciousness … and the age-old Laws to guide us'. *The Edge* magazine, 1 October 2018. See:

56 Margit Mustapa (1963), *Book of Brothers*, p.119

57 See e.g. Aartsen (2015), *Priorities for a Planet in Transition* or (2016) *Before*

Disclosure – Dispelling the Fog of Speculation.

58 Walach (2019), op cit, p.98

59 Ibid, p.69

60 Blavatsky (1888), op cit, Vol.I, p.296 (6th Adyar ed. Vol.1, p.336)

61 Paul Howard (dir., 2020), *Infinite Potential: The Life & Ideas of David Bohm*

62 Rey Hernandez, Jon Klimo, Rudy Schild (eds.; 2018), *Beyond UFOs. The Science of Consciousness and Contact with Non Human Intelligence*, Volume 1, p.103

63 Aartsen (2019), *George Adamski – The facts in context*. See: <www.the-adamski-case.nl>

64 Sylvia Cranston (1993), *HPB – The Extraordinary Life and Influence of Helena Blavatsky, Founder of the Modern Theosophical Movement*, p.xvii

65 Colombano (2018), op cit

66 Hans C. Petersen [n.d.], *Report from Europe 1963*, p.136

67 Dawna Jones, 'What is reality? Interview with Dr Ervin Laszlo', 28 November 2016. See: <www.youtube.com/watch?v=1Ke2apZ5aPk>

68 World Wildlife Fund for Nature (2020), *Covid-19, Urgent Call to Protect People and Nature*, p.25. See: <www.worldwildlife.org/publications/covid19-urgent-call-to-protect-people-and-nature>

69 Steve Curwood, interview with James Gustav Speth, 13 February 2015. See: <loe.org/shows/segments.html?programID=15-P13-00007&segmentID=6>

70 Martin Rees, 'Towards a post-human future'. Royal Institution, 16 June 2020. See: <www.rigb.org/whats-on/events-2020/june/public-towards-a-posthuman-future>

71 Adamski (1955), *Inside the Space Ships*, p.137

72 Arnold Toynbee and Daiseku Ikeda (1975), *Choose Life*, as quoted in Walach (2019), op cit, pp.91-92

73 Andrew Anthony, 'What if Covid-19 isn't our biggest threat?'. *The Observer*, 26 April 2020. See: <www.theguardian.com/science/2020/apr/26/what-if-covid-19-isnt-our-biggest-threat>

74 Wilbert B. Smith (1964), *The New Science*, 1978 reprint, p.9

75 Walach (2019), op cit, p.79

76 As quoted in Arthur I. Miller (1987), *Imagery in Scientific Thought*.

77 Adamski, letter to Emma Martinelli, 16 August 1950

"The nature of God is a circle of which the centre is everywhere, and the circumference is nowhere." –Empedocles, philosopher

2. ONENESS – THE SOURCE AND NATURE OF REALITY

The new, expanded scientific understanding of life tells us that our material reality is intricately connected with, or even dependent on, consciousness at an incipient or fundamental level. In other words, consciousness does not emerge from material or physical forms or circumstances but instead creates and informs them – consciousness (also) exists independent of a visible, physical vehicle through which it expresses on the material plane. Many examples may be given of people from all walks of life who have testified to their personal experience of being fully, often far more fully, conscious while being pronounced dead by doctors. A few poignant examples will serve to illustrate our point.

As a child Dr Murdo MacDonald-Bayne (1887-1955) saw the features of the Christ on the closed window of his room. Frightened by this, he fainted but after the experience he found he could see and hear things of which other people were unaware. He could jump from a height, stop himself in mid-air and slow down to land gently. Once, when jumping over a brook, he lost his left eye when a piece of barbed wire pierced it. Dr Mac, as he was affectionately known, had entered medical studies as a young man but found the discipline too materialistically oriented and went on to study and practise the self-healing properties of mind and body and later earned doctorates in Divinity and Philosophy.

When he was fighting in the battle of the Somme with the British Army in France in 1916, Dr MacDonald was badly wounded and left for dead on the field for four days. Years later he told one of his students about what had transpired: "During that period of time spent 'on the other side', I was instructed to return to my physical body, for my work was to tell the Truth to the world. When they came to remove the corpses for burial in a mass grave one of the ambulance men heard my groans and said, 'There's a live one here!' I was brought to the medical field station and immediately operated on. I heard distinctly the doctor mentioning that there was a possibility of losing the other eye and I decided in myself that was not going to happen. During the whole operation I was aware of what was taking place, although unable to speak, for one of the wounds was in the throat. So, out of the body, I assisted at the operation."[1]

Nearly a century later, on 10 November 2008, neuro-surgeon Eben Alexander went into a coma after contracting meningitis. Prior to his illness Dr Alexander did not believe in near-death experiences (NDEs), but what he experienced while in coma forced him to reconsider his position. Writing about his experience, he says, his memories "began in a primitive, coarse, unresponsive realm (…) from which I was rescued by a slowly spinning clear white light associated with a musical melody, that served as a portal up into rich and ultrareal realms."

He experienced "the infinite healing power of the all-loving deity at the source, whom many might label as God (or Allah, Vishnu, Jehovah, Yahweh – the names get in the way, and the conflicting details of orthodox religions obscure the reality of such an infinitely loving and creative source). God seemed too puny a little human word with much baggage, clearly failing to describe the power, majesty and awe I had witnessed. I originally referred to that deity as Om, the sound that I recalled from that realm as the resonance within infinity and eternity."

After his profound experience Dr Alexander concluded: "The conventional reductive materialist (physicalist) model embraced by many in the scientific community, including its assumption that the physical brain creates consciousness and that our human existence is birth-to-death and nothing more, is fundamentally flawed. At its core, that physicalist model intentionally ignores what I believe is the fundament of all existence – consciousness itself."[2]

As a result of such experiences, and with ongoing research into reality at the quantum level showing that it is not possible to observe a phenomenon without affecting it, science increasingly accepts the evidence of human consciousness manifesting outside the physical body in cases of out-of-body and near-death experiences, or of the consciousness of an individual 'soul' being carried over from one lifetime (incarnation) into the next.

Early research on the subject of consciousness outside the human body was done by Dr Ian Stevenson who documented *Twenty Cases Suggestive of Reincarnation* (1966).

Later, Dr Raymond Moody published his research into near-death experiences (NDEs) in his book *Life after Life* (1975), while inventor George Meek documented his research into out-of-body experiences (OBEs) and electronic voice communication with discarnate beings in *After We Die, What Then?* (1980). All the while, stories of young children displaying uncanny detailed knowledge of a previous life or death feature regularly in the media.

Taking the study of medicine as a metaphor for the study of life, in his *Science of Life* course George Adamski explained how consciousness may operate separately from the body that serves as its vehicle in the physical world. As a teacher of metaphysics in the 1930s, when one of his students would be ill and unable to attend his class, the student would later report he didn't miss the class because Adamski was teaching by their bedside: "I was there before the students giving instruction through my mind and body while at the same time my consciousness was at the bedside of the one who was sick. (…) Man is a thought manifestation of consciousness like a shadow is a manifestation of a form. So in the classroom I was manifesting as a solid form and at the bed side I was manifesting as a shadow of that form. (...) This has happened many times even outside of the classroom. When I was lecturing in Pasadena someone would ask my help for a sick friend. I would deliver the lecture on a normal basis and at the same time go to the bedside of the sick person. (…) These things did not take place during my sleeping hours but when I was active with something else. (…) But when my interest was taken up with flying saucers [i.e. from the late 1940s onward] these experiences ceased."[3]

The notion of physical reality being a mere shadow of the underlying consciousness was first introduced to the modern world in the work of Madame Blavatsky, and later elaborated on by Alice A. Bailey, and others such as Dr MacDonald, who in 1936-37 studied with several of the same wisdom teachers in Tibet, known as the Masters of Wisdom, who worked through Blavatsky and Bailey. He described his training in detail in two books, *Beyond the Himalayas* (1954) and *The Yoga of the Christ* (1956).

Adamski twice mentioned that he himself had studied in Tibet as a boy. Emphasizing he was not a Theosophist, a Rosicrucian, or anything else, in a private talk in March 1955 he stated: "I did study in Tibet when I was an eight year old boy. I took up Occult Catholicism because my father hoped I would become a priest which I decided against. I have since studied many philosophies and religions, but I didn't become associated with any one particular religion. I have taken the pearls from each and discarded the chaff."[4] In an interview with the *Los Angeles Times* in April 1934, when his Royal Order of Tibet had recently opened its Temple of Scientific Philosophy in Laguna Beach, he elaborated: "I learned great truths up there on 'the roof of the world', or rather the trick of applying age-old knowledge to daily life, to cure the body and the mind and to win mastery over self and soul."[5] Several other sources list different years and ages for his time in Tibet, which hints at the possibility that between 1900 and 1909 he spent time with his tutors in Tibet on more than one occasion.

"A man can consciously travel the Cosmos..." said Adamski in his *Science of Life* course. In fact, in private correspondence about his book *Pioneers of Space* Adamski explains in 1950 "how one may venture from one place to another, while his physical is in one place and he is in another. That is the way I have written this book. I actually have gone to the places I speak of; I actually have talked to the ones I speak of."[6] In his introductory statement in *Pioneers* Adamski said that what we are about to read, although based on the scientific facts of his day, known law and common sense logic, was "at present in the field of fiction". So when it was found that some descriptions in this book were very similar to passages in *Inside the Space Ships* (1955), that statement was quickly used to dismiss his work wholesale as science fiction. Yet, the advancing insights of science into the nature of consciousness should make it clear that his experiences were based on his higher knowledge and little known abilities.

It is in the light of this growing understanding of consciousness that the accounts of the people who claim to have been contacted by people from Mars, Venus, and other planets in our solar system in the 1950s gain renewed significance and should be reconsidered, given that scientific findings also point to a universe teeming with life, even if it isn't visible to the naked eye. As one of Mme Blavatsky's Teachers wrote to another of his students in 1882: "…whenever I speak of humanity without specifying it you must understand that I mean not humanity (...) as we see it on this speck of mud in space but the whole host already evoluted", that is, "all the intelligences that were, are or ever will be whether on our string of man-bearing planets or on any part or portion of our solar system".[7]

40

To better understand how life is One and everything is interconnected and interdependent, it might help if we compare how the teaching of the ageless wisdom and science look at the origin of existence. This will necessarily be limited to general observations, but if correspondences are found at a fundamental level, that should give us reason to postpone our judgement on the differences in details.

If everything we know and see, ourselves included, originates from the same matter-generating field of universal mind or consciousness, this means there is an underlying unity that expresses at an endless variety of levels – indeed from quanta to galactic superclusters. "The one Life, which we call God, expresses itself through a myriad forces and forms," according to the late esotericist Benjamin Creme. He says: "The awareness of God, the knowledge of the nature of God, comes to us as we expand our consciousness, and therefore our sensitivity, to the various forces which together make God. At the centre of the galaxy there is a Being Who has that consciousness on a galactic level, which is so far above that of a solar system that one cannot even imagine it." This Being is referred to in the wisdom teachings as "The One About Whom Naught May be Said". Not, says Creme, "because it would be irreverent to say anything about that Being, but because there is nothing we can say, there is nothing we could imagine to say about the Being at the centre of our galaxy."[8]

How the manifestation of boundless diversity has come about from what science ('Big Bang'), religion ('God'), and visitors from space ('Infinite Father') agree must be One Source, is beyond my puny understanding or capacity to explain or even summarize from the wisdom teaching, which often refers to the originating Cause as the Absolute. In so many words, theoretical physicist Michio Kaku admits that mainstream science can't either: "The strength and weakness of physicists is that we believe in what we can measure. And if we can't measure it, then we say it probably doesn't exist. And that closes us off to an enormous amount of phenomena that we may not be able to measure because they only happened once. For example, the Big Bang… That's one reason why they scoffed at higher dimensions for so many years. Now we realize there is no alternative…"[9]

Systems science, in turn, offers a beginning of understanding in the words of Ervin Laszlo: "The map of consciousness and the map of the cosmos maps the same reality, it only focuses on different manifestations of it."[10] Professor Laszlo draws from the work of physicist David Bohm who stated that underneath the macrocosm and the microcosm is the 'implicate order'. This is not a set of objects, explains theoretical physicist Dr David Peat, "but a process of constant movement, constant unfolding and enfolding. So the explicate order, the three-dimensional world, comes out of the implicate order. Therefore, each part of the cosmos contains the whole of the universe and unfolds into our perception of reality."[11]

According to Blavatsky, "The appearance and dis-appearance of the Universe are pictured as an outbreathing and inbreathing" of the Great Breath, "which is eternal and which, being motion, is one of the three aspects of the Absolute – Abstract Space and Duration being the other two."[12] Thus, "The appearance and disappearance of worlds is like a regular tidal ebb of flux and reflux."[13]

The two volumes of *The Secret Doctrine* that Mme Blavatsky managed to finish of her major opus before her death in 1891 describe and document the evolution of the cosmos (*Cosmogenesis*) and the evolution of man (*Anthropogenesis*), based on ancient records and rare texts that support her uniquely incisive interpretation of the previously esoteric knowledge behind the exoteric representation in religious and other teachings. These she exhibited in the context of the science of her day – boldly exposing the inconsistencies in its reasoning and deductions where she encountered them, while crediting scientific hypotheses or findings that were mostly unknown or being ignored.

Taking up one full volume of *The Secret Doctrine*, the endless process of the ideation and creation of the physical universe through the originating Mind or consciousness 'involving' itself in denser and denser projections of itself and, from the lowest point of the cycle, evolving back towards its original state through self-aware individualized units of consciousness, is described in unprecedented and unparalleled detail. Unparalleled, that is, in the writings that are now accessible to our modern, inquisitive minds which are for the most part still disconnected from the natural order of life, the original state of oneness that is obscured from our view in the trappings of our dense-physical existence.

Some may see the recurring cycle of this involution into 'matter' and evolution back into the original 'spirit' reflected in the universe expanding from the Big Bang and eventually returning into the oblivion of black holes, and that may not even be so far off the esoteric mark. In Blavatsky's words: "DARKNESS [the Absolute] is the one true actuality, the basis and the root of Light, without which the latter could never manifest itself, nor even exist. Light is Matter and DARKNESS pure Spirit. Darkness, in its radical, metaphysical basis, is subjective and absolute Light... Even in the mind-baffling and science-harassing *Genesis* [2-3], light is created out of darkness (…) and not *vice versa*."[14]

Elsewhere Blavatsky elaborates: "We believe in no *creation*, but in the periodical and consecutive appearances of the universe from the subjective on to the objective plane of being, at regular intervals of time, covering periods of immense duration. (…) As the sun arises every morning on our *objective* horizon out of its (to us) *subjective* and antipodal space, so does the Universe emerge periodically on the plane of objectivity…"

While science would call this process evolution, and to the pre-Christian philosophers and Orientalists it is known as emanation, to the wisdom tradition it is, in Blavatsky's words, "the only universal and eternal *reality* casting a periodical reflection of *itself* on the infinite Spatial depths. This reflection, which you regard as the objective *material* universe, we consider as a temporary *illusion* and nothing else. That alone which is eternal is *real*."[15]

In February 2020 a group of quantum gravity researchers published a commendable new interpretation of quantum mechanics in a paper "demanding origin stories for every-thing, even those things that are supposedly fundamental". They come remarkably close to Blavatsky's explanation of the ultimate origin as "eternal reality casting a periodical reflection of itself" when they propose that consciousness and the physical universe, are "information in a simulation run in the mind of the emergent panconscious universe – the self-emergent substrate as a strange loop."[16] However, where they suggest that "hacking of evolutionary biology (...) is likely to allow rapid evolution of consciousness in the future"[17] they seem to return to the notion that consciousness is not at the root, but emerges from biological organisms.

In *Origin and Evolution of Mankind* (1986) Dr Birendra Kumar attempts an admirable summary of Mme Blavatsky's teaching on the manifestation of the cosmos. The Universe – the handicraft of a Supreme Absolute Being, the One Reality – "is pervaded by duality where spirit is linked to matter by an unknown force which the occultists have called 'FOHAT'. Fohat is the bridge by which the 'Ideas' existing in the 'Divine Thought' are impressed on cosmic substance as the Laws of Nature".

In this context, it is noteworthy that Ervin Laszlo calls the laws of nature "instructions, precise algorithms for the evolution of coherent systems in the spacetime domain of our universe".[18] He also addresses the question of how abstract information (such as 'Ideas' of manifestation) could generate physical events (via the laws of Nature) and says unreservedly: "Without recognizing a dimension that is part of the cosmos (but is not the dimension of spacetime) this question cannot be answered."[19] Calling this the "deep dimension", Laszlo in fact says intelligence (Blavatsky's 'Divine Thought') permeates the universe.

Dr Kumar's summary continues: "The entire Universe and its fragments, the Solar Systems, our Solar System, Moon, Planets, Stars and the Earth is made of original ultimate root matter. This is described as Koilon in occult chemistry and aether of space by Science... Koilon is derived from Greek Koilos, i.e. *hollow*." Yet, being denser than anything conceived of by the human mind, the Supreme Being activated the Koilon by blowing a mighty breath into them and an "infinite number of bubbles were created in space. These bubbles in Koilon constitute the ultimate atoms out of which all matter was generated."[20] Madame Blavatsky described the force which formed the bubbles as "Fohat digging holes in space".

Here we should note that the concept of 'ether' in science was introduced in the 19th century by French physicist Jacques Fresnel as the invisible element that fills all space. When early 20th-century experiments failed to prove Fresnel's hypothesis, science simply dismissed the notion of ether. But, as professor Laszlo points out, the concept of this dimension beyond spacetime (i.e. the observable universe) was reintroduced to physics when it was shown that "the energy levels of the universe exceed the known dimensions of spacetime. Spacetime could not be the entire reality of the universe."[21] If astrophysicists refer to this unknown 'dimension' as "dark energy", esotericists call it 'etheric' or subtle matter.

How could the concept of Fohat be reconciled with the insights from mainstream science? We should not assume that the Perimeter Institute for Theoretical Physics in Canada keeps a copy of Blavatsky's *Secret Doctrine* on their desks for reference, as Albert Einstein did.[22] Nevertheless, in 2014 it proposed a hypothesis of the early formation of the universe that evokes a very similar image of its manifestation. In a report for the *SciTechDaily* website, associate faculty member of the Institute Matthew Johnson writes that in the beginning was the vacuum (cf Blavatsky's hollow *Koilon*): "The vacuum simmered with energy (variously called dark energy, vacuum energy, the inflation field, or the Higgs field). Like water in a pot, this high energy began to evaporate – bubbles formed. Each bubble contained another vacuum, whose energy was lower, but still not nothing. This energy drove the bubbles to expand. Inevitably, some bubbles bumped into each other. It's possible some produced secondary bubbles. Maybe the bubbles were rare and far apart; maybe they were packed close as foam. But here's the thing: each of these bubbles was a universe. In this picture, our universe is one bubble in a frothy sea of bubble universes."[23] The similarity with the notion of primordial matter originating from bubbles in the theosophical description above is striking.

The concept of reality according to systems science sounds very similar in the words of Ervin Lazlo: "The manifest world is a set of clusters of coordinated vibration in the excited [i.e. the manifested] state of the cosmos." The fact that these clusters of vibration are coordinated indicates that the manifestation of reality is not a random occurrence: "The clusters are 'in-formed' by a factor we identify as an underlying cosmic intelligence."[24] It is remarkable to read how closely George Adamski's teaching resembles this: "Consciousness is the father and mother of all form creation which conceives and gives birth to the various forms. And within it is the blueprint or memory which is ever present..."[25]

"In this vast universe space is like the ocean and the planets within it are comparable to the various strata of pressure within the vast ocean," mused George Adamski in the foreword to his book *Pioneers of Space*, in which he described his out-of-body excursions to other planets. "While this is at present in the field of fiction, the advance of science is so rapid that it will not be long before all of this will become a reality."[26] Although the recent findings as summarized by Ervin Laszlo strongly support this statement, the real significance of Adamski's first book about space travel, however, for me lies in the fact that it already exudes a profound sense of oneness throughout his account of visiting other planets and his meetings with highly evolved beings there.

Harking back to Blavatsky's work, and forward to the advancing insights of 21st century science of the universe emerging as an 'idea' from 'divine thought' (Blavatsky) or cosmic intelligence (Laszlo) before being brought into manifestation, in *Pioneers* Adamski describes how he and his travel companions were given an explanation of how ideas or thoughts pass through space: "The invisible manifestations are prior to form expression. In the beginning, form is preceded by what might be called an imaginary complete picture of what the form is to be. Once this is gathered in completion, slow-up action of all these minute units making up the imaginary picture starts taking place. As this slow-up action continues, the form gradually begins to come into visibility as a manifestation, for the difference between the visible and invisible is speed, or lower and higher vibrations."[27]

"Separately they are almost like pieces of a jigsaw puzzle, but when they get together they form a complete picture of an idea. (...) So the invisible thought or picture may be called imaginary but it is real and can be brought into material or solidified manifestation. Of course back of this life force manifestation is its perception and birth. Back of all forms lies the Fatherly intelligence of the Infinite Kingdom, called God. Beyond this, man cannot go."[28]

As a student of the Ageless Wisdom teaching I have often referenced Benjamin Creme's assertion that "All the planets of our system are inhabited, but if you were to go to Mars or Venus you would see nobody because they are in physical bodies of etheric matter"[29], and presented supporting evidence for his suggestion that various – what are often considered 'fringe' – discoveries actually point to the reality of the etheric (subtle) planes of matter. Yet, from the esoteric perspective there is more to the notion of life on other planets being invisible to our dense-physical sight than planes of matter above the solid, liquid and gaseous states. The idea of 'different dimensions' that many scientists and seekers alike speculate about seems to be rooted in the composition of the manifested cosmos.

While it would take up too much space, and too much of my ability, to provide a comprehensive exposition of the wisdom teaching on this subject, some references may give the reader a hint at the underlying reality, especially given the fact that its fundamental premises are increasingly embraced by 21st century science.

For instance, Mme Blavatsky explains: "Whether by radiation or emanation (…) the universe passes out of its homogeneous subjectivity on the first plane of manifestation, of which planes there are seven, we are taught. With each plane it becomes more dense and material until it reaches this, our plane, on which the only world approximately known and understood in its physical composition by Science, is the planetary or Solar system [which], like every other such system in the millions of others in Cosmos, and even our Earth has its own programme of manifestations differing from the respective programmes of all others."[30]

In other words, each solar system, and every individual planet in a solar system, has its own purpose to fulfil on its evolutionary journey. It would seem this is what Canadian researcher and contactee Wilbert Smith referred to when he said: "These people [from elsewhere] tell us of a magnificent cosmic plan, of which we are a part, which transcends the lifetime of a single person or a nation, or a civilization, or even a planet or solar system. We are not merely told that there is something beyond our immediate experience; we are told what it is, why it is, and our relation thereto."[31]

Likewise, a "profound space teacher" told US contactee Howard Menger: "Man continually seeks his Source, the Supreme Consciousness – and those great men of your holy writings – who touched upon this Source, discovered a divine plan for all mankind, one rooted in love; for the Supreme Consciousness is love. We are dedicated emissaries of this divine plan, to your planet, to those with an evolved insight, for those are the ones who will receive us."[32]

The "passing out of its homogeneous subjectivity" in the case of a solar system and its planets is a process that happens in seven stages, each of which constitutes an entirely different 'layer' of space and which, in Blavatsky's description, "exists in nature outside of our normal mentality or consciousness, outside of our three dimensional space, and outside of our division of time."

What's more, each of these seven fundamental planes in space "has its own objectivity and subjectivity, its own space and time, its own consciousness and set of senses", as different from our consciousness and senses as those we experience in our dreams, when events that span several days or even weeks pass in an instant. Put in the simplest of terms, then, there exist seven planes of being in cosmos and therefore "seven states of consciousness in which man can live, think, remember and have his being."[33] Here, I believe, we find the original thought that informs modern speculations about extraterrestrial visitors originating from other dimensions, or scientific hypotheses about the 'multiverse'.

These seven planes come into being as a result of an outbreathing of the creative aspect of the *Logos* (Greek: 'word') – which we find echoed in the opening passage of the Gospel according to John ("In the beginning was the Word..."), and in the sacred sound in Hinduism ("Aum"; or the 'Om' experienced by Dr Alexander) that brings everything into manifestation through a process of gradual descent from the highest level of abstraction.

As a side note, when the Brazilian physicist Aladino Félix (wrriting as Dino Kraspedon) was contacted by a visitor from space in 1952, the same year that George Adamski reported his first physical contacts, he was told: "I know that the Universal Life has an enchantment beyond words; as though it were some mysterious song, sung by some immortal Being, whose voice brings worlds into being, then destroys them to re-create them. At the command 'Talitha Cumi' the universes stream forth again."[34]

According to one of Mme Blavatsky's Teachers, the

51

sevenfold manifestation of cosmos also applies to planets. In fact, she says, we should imagine each planet as a chain of seven beads representing the globes on which the 'Breath of Life' from the planetary Logos completes seven rounds in succession, as it spirals downward from the first globe at the top of the chain before passing on to the next. Our planet is only the fourth globe, and therefore the lowest – and densest – in the Earth chain, on which the human kingdom itself is now at the halfway point through the fourth cycle (round). From this point onward the manifestation of the natural kingdoms, including the human, on subsequent globes will be progressively less physical, as the originating force draws the resulting self-conscious units back to the Source.[35]

In his meeting with "one of the most advanced beings from another planet" whom Howard Menger met he is told something that hints at the same notion of evolutionary cycles: "Everything created has a consciousness, and the consciousness evolves to the soul point, where it expresses in higher forms, man being the highest. This doesn't necessarily mean one evolves from the other, but there are many cycles of evolution."[36]

The fact that our probes don't find signs of life on the other planets in our solar system may be due not only to life manifesting there on the subtle (etheric) planes of matter as it does in the case of most planets, but also to the fact that the life stream of a planetary manifestation may be on a different globe in that planet's chain than our globe in the Earth chain, and is therefore beyond not only our vision, but even beyond our cognitive grasp.

Clearly, there is no hard evidence yet for this process of

the manifestation of life, nor does science as such accept the circumstantial evidence that is presented in the wisdom teachings about present-day humanity on Earth being the evolutionary outcome of previous cycles of the human (or proto-human) evolution, successively known as the Polarean, Hyperborean, Lemurian, and Atlantean. In *The Secret Doctrine*, however, Mme Blavatsky provides much evidence and many reasons, often in the form of discrepancies in existing scientific explanations, why such previous cycles would solve many of the remaining anomalies in mainstream geology, palaeontology and anthropology.

Yet, if nothing else, the growing convergence of science and the wisdom teachings presented in this and the previous chapter shows us that there are no 'hard-and-fasts' when it comes to understanding the universe, consciousness, life, and reality. Science and scientific insights evolve as our understanding expands with our consciousness, and what was once esoteric or sequestered in the realm of the 'mystical' is gradually being entertained in scientific hypotheses.

George Adamski once made a statement that could serve as an explanation: "There is continuous interblending between the visible and the invisible – the high vibration and the low – but never a break... When we speak of matter we are speaking of the spiritual in a low state of manifestation... The trouble with the metaphysical setup is that everything in the invisible is labelled 'spiritual' while in the visible it is labelled 'material', but in truth there is neither spiritual nor material – it is all the same."[37]

Dr Steve Taylor, lecturer in psychology at the Leeds Beckett University, UK said something very similar: "You could say that matter is the external manifestation of spirit, while mind is its internal manifestation."[38] So, when Carl Sagan said: "We're made of star stuff. We are a way for the cosmos to know itself", he basically rephrased from the 'low' or material state of manifestation what Jesus of Nazareth said from the 'high' or spiritual state, when he proclaimed: "I and the Father are One".

It is interesting to note, based on these examples, that although the scientist and the student of the wisdom teachings arrive at strikingly similar insights, they approach reality from opposite ends of the spectrum. As Dr Kumar writes: "Viewing from above the occultists have been able to trace the descent of matter and energy into their respective channels. The scientists face the difficulty of tracing the ascent of matter after they reach the dead end of the alley of their knowledge and technique. They have to wait till new lenses for higher observations are discovered."[39]

Or, as Mme Blavatsky put it: "The methods used by our scholars and students of the psycho-spiritual sciences do not differ from those of students of the natural and physical sciences... Only our fields of research are on two different planes, and our instruments are made by no human hands, for which reason perchance they are only the more reliable."[40]

Our examples and references show that as these metaphorical 'new lenses' are discovered, science comes ever closer to the postulates of the wisdom tradition. While studying with the Masters of Wisdom in 1936, Dr MacDonald-Bayne was equally told: "From the visible to the invisible and beyond, there is no division, and from beyond the invisible to the visible there is no separation, and in and through, supporting this change, is the changeless basic substance that remains stable always. And beyond and within is the Creativeness that uses this substance to create form. (…) Now, what we do not know is the Uncreated which alone is creative. And this Uncreated is within you; you can discern all that is relative to It, but you cannot discern what It is itself because It will always be discerning that which is external to Itself."[41]

Recognizing consciousness as an aspect of reality quite distinct from the physical aspect, therefore, will provide not only scientists but UFO researchers as well with a lens that allows them to penetrate much farther into the mysteries they are probing and seek to resolve.

Indeed, the apparent elusiveness of the UFOs seems a fitting metaphor for our struggle to understand life, with mainstream science still chasing the ultimate building block of life to firmly land the feet of its understanding on. More attentive scientists have found that this search has revealed there are no such building blocks, at least not in any form of permanency.

The deeper science looks into the building blocks of life, beyond chromosomes and enzymes, it finds electrons, protons, quarks et cetera, and it can no longer make the distinction between living or dead matter. This, too, is confirmed in the wisdom teachings, when Benjamin Creme said: "There is no such thing as dead matter; all matter is conscious. It is the consciousness of the atom, the desire principle inherent in every atom in the universe, that drove the fish out of the sea and onto dry land, to become a mammal and eventually a human being."[42]

British wisdom student and author Vera Stanley Alder

says: "It must be always remembered that the Ageless Wisdom teaches that the universal soul-life which pulsates throughout all the atmosphere is very eager and ready to ensoul and *entify* (cause to become a separate entity) any manifestation on planes lower than its own when it has tended to individualize itself... Therefore every possible aspect of life is more likely to be entified than not... This means that all things are organic, or a part of some organism, and that therefore there is actually nothing inorganic in the world at all."[43]

As to the nature of consciousness, Dr Mac was told: "You cannot tell what your consciousness is – try to see if you can, and you will find that consciousness is always discerning what is relative to itself. It cannot turn back upon itself. But when all the relative is understood and known, then the Unknown can be experienced. It cannot be known, for the known cannot know the Unknown; therefore the known is not the Real, the Real is the Unknown, the Unknowable."[44]

In *The Universe, Life and Everything*, titled after the third volume in the *Hitchhikers' Guide to the Galaxy* series, scholars from a variety of backgrounds, among whom Alexander Wendt, discuss how the traditional understanding of our world is challenged by developments in (quantum) physics, and detail our inability to explain a complex phenomenon such as consciousness. The dialogues, conducted by professor of Neuroscience Sarah Durston and psychotherapist Ton Baggerman, show remarkable agreement "that there is a shift occurring in the way we understand reality, precisely because our current world view is reaching its limits. People are beginning to understand the world and themselves as interconnected, rather than as individual entities bound together only by cause and effect."[45]

We are reminded of the words of Mexican contactee Carlos Díaz, who was contacted in March 1981 by what he describes as human-looking extraterrestrials, and whose contacts continued for many years.[46] According to Mr Díaz his contacts imbued him with an awareness of the interconnectedness of all life, strongly reminiscent of the messages given to other contactees: "I have experienced that there is a wonderful interaction between all living things. But this interaction has been disturbed. Every individual, every species is an important part of this interaction."[47]

French contactee Pierre Monnet, too, asserted that the way we think and feel has a direct impact on our environment: "In the majority of cases, cataclysms, even natural ones, are caused by the negative thoughts of the beings who inhabit the planet." He went so far as to claim that the majority of earthquakes "result from the negativity of human thoughts. I have already said: negative thoughts of humans are more important than positive thoughts. These negative thoughts group together and take on an unexpected force, forming an aggregate that feeds the collective unconscious, crystallizing and acting on matter by boomerang effect..."[48]

In an article in 2007 the Master of Wisdom who worked via UK esotericist Benjamin Creme explains this interrelationship as follows when he writes about natural disasters: "It is not a question of God's love failing humanity but of seismic pressure which must be released. What, we may ask, causes seismic pressure to grow to such a destructive extent? Elemental Devas [the building forces of nature] oversee the mechanism by which these colossal energies act or are modified. The Earth is a living Entity and responds to the impact of these forces in various ways. One major source of impact comes directly from humanity. As humanity, in its usual competitive way, creates tension through wars, and political and economic crisis – that is, when we are out of equilibrium – so too do the Devic lives go out of equilibrium. The inevitable result is earthquakes, volcanic eruptions and tsunamis. We are responsible."[49]

A case in point may be found in the Covid-19 pandemic that brought the world as we knew it to its knees in February 2020. Since the 1960s humanity has ignored the warnings of scientists that our lifestyle of exhaustive materialism, pervasive pollution, and wasteful overconsumption threatens the living organism that is planet Earth, while also ignoring the plight of millions who starve and die for lack of basic necessities. At the same time our leaders act as if our socio-economic structures, that only benefit the few at the expense of the many, are an immutable fact of nature. And while many scientists have blamed bats and exotic 'wet' animal markets as the source of the virus, one commentator says "the real blame lies elsewhere – with the human assaults on the environment."[50] In an interview titled 'We did it to ourselves' the godfather of biodiversity Thomas Lovejoy, who coined the very term, agrees: "This pandemic is the consequence of our persistent and excessive intrusion in nature..."[51]. For this reason, professor Laszlo says, "we have to recover our sense of being part of that whole system and act in a way that we ... become a constructive part of it, which we have been."[52]

In recent years more and more scientists and medical

professionals seem to be breaking the academic 'taboo' of seeing life, and consciousness, as not a product of chemical processes between material components in conducive circumstances, but rather as the origin or foundation of the material world.

This can even be seen in the various reconsiderations of long-held beliefs. For instance, in his book *The Evolutionary Human. How Darwin got it wrong*, author Richard Barrett proposes a new theory of evolution because, he says, Darwin left out consciousness. Barrett proposes that evolution is the result of adaptive thinking and intelligence over time, to account for consciousness as an inherent quality: "Every living entity has a mind, including the cells in our body, and every mind is primarily focused on staying alive and procreating. Enabling an entity to stay alive is the fundamental purpose of evolutionary intelligence. If an entity cannot stay alive, it cannot procreate. Without the *will to live and procreate*, evolution would never have happened. The Earth would be a dead planet. This raises an important question: 'Where does evolutionary intelligence – the intelligence that is required to stay alive and support procreation (…) come from?' "[53]

Ironically, an in-depth review of Charles Darwin's work yields results that indicate he may not have been wrong, but rather that the interpretation of his work has been incomplete or biased, mostly focused on his phrase "survival of the fittest" with which he tried to explain the complex relation between the two principles of evolution in his book *On the Origin of Species*: natural selection and variation. Dr Steve Taylor found that Darwin, in fact, didn't see evolution as competition between rivals: "Darwin was fully aware of the cooperative aspects of evolution, and gave many examples of them in *On the Origin of Species*. (…) Darwin did use the term 'struggle for existence', but stated that he used the term 'in a large and metaphorical sense, including dependence of one being on another'."[54]

Systems scientist David Loye points out that the term was quickly appropriated and misused by empire-builders who "hailed 'survival of the fittest' for scientific proof of their right to seize and rule the 'backward' people of this earth. (…) The strong must prevail over the weak." However, Dr Loye's research shows, Darwin's theory was not a mechanistic process towards physical perfection, but heavily infused with the notion "that well-being of others was an equally, and in many cases, even more powerful drive".j [55] He found that 'survival of the fittest' occurs only twice in Darwin's next book, *The Descent of Man*, against writing 95 times of love, 90 times of moral sensitivity, and 200 times of mind and brain. Writing on The Darwin Project website, Dr Loye says: "Suppression over 100 years of the *real* Darwin has led to the social, political, economic, scientific, educational, moral and spiritual mess we are in today." Indeed, it is as if Darwin himself hints at the evolution of consciousness when Dr Loye quotes him in his book *Rediscovering Darwin. The Rest of Darwin's Theory and Why We Need It Today*: "…a belief in all-pervading spiritual agencies seems to be universal; and apparently follows from a considerable advance in man's reason, and from a still greater advance in his faculties of imagination, curiosity and wonder."[56]

It is interesting to see how objections to Darwin's theory on evolution may be countered by arguments based on unknown aspects of Darwin's work itself. Equally interesting is how a Master of Wisdom who worked with Benjamin Creme addresses the shortcomings of both the Darwinian concept of evolution and the religious notion of creation: "[E]volutionists and the creationists are really arguing at cross-purposes; both, in their limited way are right. (…) The creationist is at pains to emphasise that 'Man' was made by God, in 'God's own image', and so owes nothing to evolution. To such, Darwin and those who follow him are missing the point about Man: that he is a spiritual being, of divine heritage... From Our point of understanding the scientists of today, the evolutionists, are undoubtedly correct in their analysis of Man's development from the animal kingdom. (…) That, however, does not make us animals. Darwin, and those who correctly followed his thought, describes only the outer, physical development of Man, largely ignoring that we are all engaged in the development of consciousness. The human body has all but reached its completeness: there remains little further to be achieved. From the standpoint of consciousness, however, man has scarcely taken the first steps towards a flowering which will prove that man is indeed divine, a Soul in incarnation."[57]

The evolution of consciousness, like the evolution of the physical form from the primordial soup, is not random, mechanistic, or without direction; it has been shown to move toward ever greater synthesis and unity. This is also confirmed in the accounts of the 1950s (and many later) contactees where they write about the compassionate and wise counsel which the visitors shared for the benefit of humanity.

Of these, George Adamski stands out as the first to

write and speak out publicly about his contacts for his ground-breaking travelogue *Pioneers of Space* in 1949. In it he describes what Dr Walach refers to as an out-of-body experience of non-local consciousness, many examples of which he documented in his Galileo Commission Report. When Adamski explains how we may learn to direct our consciousness elsewhere, he explicitly distinguishes between his travels in consciousness and "my trips in space craft taken bodily".[58] In a letter to a reader Adamski had clarified in 1950: "Yes, one may travel at will [to] any place in the universe without taking his physical body since the physical is not man, but rather the house of man. Man himself is what is termed the spirit. (…) The spirit is nothing more nor less than the conscious consciousness [self-consciousness] which is called man and is no different than the conscious consciousness known as the Divine Father of the universe – and it is everywhere. (…) the consciousness had to be before the building, or house within I live, could have been built. And that is also true of the physical body."[59]

Another contactee who was shown the planet of his contacts from space in an out-of-body experience was Frenchman Pierre Monnet (1932-2009). Virtually unknown in the English-speaking world, he was a conscript at the time of his first contact experience in 1951 in Courthézon, near Avignon, France. Neither of his two books, *Les Extraterrestres m'ont dit…* ('Extraterrestrials told me…', 1978) and *Contacts d'Outre-Espace* ('Contacts from Outer Space', 1994), have been translated in English, which might explain why Mr Monnet's story is not more widely known. In an interview in 2006 he describes his excursion as a "telepathic journey" because he says he lacks the terminology to describe the 'mechanics'.[60]

When we understand, from science and the wisdom-religion ('religion' in the radical meaning as a technique to reconnect), that consciousness is fundamental to physical reality, and only depends on a physical form to express and experience life in any particular stratum of the ocean of consciousness – any of Blavatsky's seven 'layers' of space – we can also see how our individual consciousness may be directed somewhere outside the physical body and regardless of our everyday experience of dimensions.

As a student of the wisdom teachings, British portrait artist Vera Stanley Alder describes in her autobiography how in 1942 she was taught to consciously leave her physical body and was shown lives expressing in both the macro-cosmos and the micro-cosmos. On one such expedition she was shown a cell in a human body: "I watched expectantly as the great roundish cells swelled upwards until we could only see the one nearest to us, which lay shackled to the ground and to its neighbours by a tangle of jungle-like rubbery tree-trunks. We passed between this forest of gleaming, shifting trunks and entered the cell through one of the cavernous openings which pierced its sides. It seemed like entering a great domed Olympia, into whose interior was packed an amount of varied activity and a large number of strange forms. (…) We passed long enormous sausage-shaped creatures which swelled and curved and appeared to be full of life. There were also great round bodies with speckled surfaces, and smaller beautiful ones like jelly-fish. Tubes and strands of all sizes threaded their way in and out. Finally we approached a large collection of big banana-shaped creatures curled around each other tightly.

In between them nestled solid-looking globes of pink, dark red and brown. 'This must be the nucleus of the cell!' " The author was then given the opportunity to commune with the cell and when she asked if he was content to be a cell the "clustering bodies shivered as I spoke and I could have sworn it was in eagerness to answer me."[61]

Anyone who wonders if such an astonishing account should even be considered in earnest, should ask the same with regard to the experience described by a Nobel Prize laureate whom Dr Harald Walach quotes as a remarkable example of 'radical introspection' in his report for the Galileo Commission. According to her biographer, cytogeneticist Barbara McClintock described how she "merged" and became one with her cells in very similar terms: "…when I was really working with them I wasn't outside, I was down there. I was part of the system. I was right down there with them, and everything got big. I even was able to see the internal parts of the chromosomes – actually everything was there. It surprised me because I actually felt as if I were right down there and these were my friends."[62]

Hinting at similar 'movements' in consciousness, in his *Science of Life* course, in a lesson titled 'The Relationship of All Creation', George Adamski writes about DNA as "tiny memory molecules [that] are actually conscious entities capable of maintaining the form and guiding the mind if the ego allows it to do so. (…) This proves that the human mind once properly schooled can commune with all forms in nature… This discovery has been made with the help of our space brothers many years ago, but not until now do our scientists realize its value and potential."[63]

Adamski's out-of-body experiences took him to the Moon, Mars and Venus, where he encountered people who look very similar to the human species on Earth. He describes his journey in terms of a physical one, where he boards a spaceship to the Moon with three companions: the pilot Bob, co-pilot Johnny, himself the navigator, and a scientist whom he calls Dr Johnston – a thinly veiled reference, I suspect, to Dr Josef Johnson, who worked as an astronomer at the Palomar Observatory in California, where Edwin Hubble had just exposed the first image with the Hale Telescope on 24 January 1949. The son of George's first benefactor in the 1930s, Lalita (Maud) Johnson, who was a teacher of metaphysics herself, Dr Johnson was a friend of Adamski's. So when Adamski suggests he didn't go on his out-of-body journey alone, it is not inconceivable that Dr Johnson was indeed capable of out-of-body travel himself and joined Adamski on some of his expeditions.

On the Moon, the party are welcomed and shown around by people from different planets who were based there, and given a tour of a ship from Jupiter. They are then invited to visit Mars, where they are taken by a Martian ship. After a tour of the planet, another ship takes them to Venus before they are returned to the Moon to board their own ship back to Earth. Given Adamski's own private admissions, it is reasonable to assume that he chose to present his journey as if it were physical to make his experiences easier to accept and digest for his readers.

On the Moon he was told: "Many centuries ago (…) we established this trade center here on your Moon, and as you see, planes from many planets in our solar system travel to and from here regularly. This was done with the hope that some day men on Earth would so develop that they could see these ships coming in and out from their own Moon and come here to investigate, out of curiosity. Your race of people have developed machines through which they have seen our ships flying, but they do not understand and for the most part do not even believe what they are seeing."[64] While he would later become world famous for the most detailed photos of flying saucers – and the subject of countless unfounded or refuted accusations of fraud – Adamski took his first photographs through his telescope of space ships as they passed across the face of the moon in 1947.

The people he encountered on his journey differed from terrestrial humanity only in their profound understanding of Oneness as the natural order of life that our science is only now hesitantly beginning to perceive and embrace: "Judging by the appearances and behaviour of the pilot and the crew of this ship from Mars, we felt like little children. Emanating from them towards us is no feeling of personal pride or attainment, neither is there the slightest indication of hostility but rather a feeling of warm friendship and joy that we have made the trip here."[65]

"Life [on Venus] is so different, so beautiful, harmonious, and joyous, and the individuals so kind, the word refinement doesn't do them justice. It makes us feel that we haven't yet begun to live."[66]

Earlier Adamski and company were told: "The Earth too may in time be blessed to know and enjoy these blessings of our Father, when our brothers in the Earthly home have learned this great lesson of brotherly state and the love of the Father that put them there. Then they will also possess the wonders that we have upon Mars.

"But as scientists, as servants of the Almighty, we never impose upon others that which they may not want. The minds of the Earthly brothers must be renewed if these wonders are to be known by them. At present their minds are not only on conquering one another, but even on conquering others like those of us and many within the kingdom of our Father, including his kingdom. Yet it is not these things they must conquer. It is themselves individually. They must conquer there in order to realize that each is his brother's keeper."[67]

The Martian scientist who addressed the visitors echoes here the age-old axiom that to understand the world we live in, to understand life, we must first understand ourselves. Understanding ourselves as an individualized aspect of the underlying oneness allows us to master our lower nature and its bodies of expression – mental, emotional, physical – to eventually become a Master of Wisdom, reconnected in full conscious awareness with the source of our being, and fully able to manifest our understanding and awareness in right relationship with the Source, with ourselves, and our surroundings. But even before we reach that stage, we all experience moments in which we feel profoundly connected with life, the universe and everything around us.

Sometimes these experiences are induced through the agency of a higher, more evolved consciousness pouring forth its energy among those in its presence, who experience this as a 'blessing'. An example of such an experience was described by Dr Mac when, after three days in Kalimpong, east of Darjeeling, in 1936 he was finally approached by the stranger who would guide him into Tibet: "He put his right hand on my left shoulder and I felt as if I was charged with electricity."[68]

In *Pioneers of Space* George Adamski describes several instances where he and his party felt 'blessed' in this way, for instance after hearing the Martian scientist speak as quoted above: "…after hearing that talk from a great scientist, we did not know whether we were eating or dreaming for it had caused something to happen within us which made us feel very humble. At the same time we felt quite small, for within ourselves we knew that we had come to the Moon towards the possibility of future exploration. Now we realized that we as Earth men understood but little in the channel of life…"[69] Later in their journey, when the planetary heads of Venus, Saturn and Jupiter address the terrestrial explorers after a feast in their honour, Adamski describes a similar experience even more explicitly: "… the three [leaders] came around the table and placed their hands upon us and blessed us. We could feel the power go right through each one of us."[70]

When he attended an interplanetary council that was held on 29 April 1962 on Saturn, he experienced this yet more profoundly: "During the eighteen hours of the assembly (earth time) it seemed to me that I no longer had a mind of my own nor did I feel as a person, but rather as an important part that fitted into a complete being expressing itself in the highest degree of its knowledge with Cosmic feeling." This experience of oneness was so strong that he had difficulty adjusting to 'normal' life once back on Earth, to the point that he said: "It is harder than anyone realizes to adjust to earthly conditions after such a trip. I don't want another like it and I surely don't want to take this body to another planet in order to go on living."[71] In the second part of his report, he explained: "While we on the inside of the ship remained as normal as you are when reading his, our physical bodies experienced a sensation of lightness and a feeling that was indescribable – of eternal well-being. There was no awareness of distance away from the earth or strangeness and my mind gave me the sensation of being cared for by delicate hands. Yet I was fully aware of a change taking place within my body and was later informed that the molecules of my body caused the feeling of oneness."[72]

Many other contactees have described similar experiences. One such can be found in Giorgio Dibitonto's account of him and his companions being invited on board a space craft in 1980: "I was aware of invisible worlds and the brothers that inhabited them. I sensed a love which embraced all Creation, and a longing which drew my whole being toward the Father. It was as if showers of blessings were pouring down on us from above, suffusing us with deep feelings of cleanliness and release. Now it was the figure of the Lord that shone with a brightness that rivaled the sun, and that fine golden light bestowed on everyone present a feeling of ecstasy and inner wholeness. In this house of living light, I was aware of entire universes. I knew that man had no limitation in the dimension of the spirit, and, at the same time, I felt enveloped in perfect peace and melted with quiet rapture. When all this was at its highest point (…) a sublime fire took possession of every living being present. It was like a golden cord, from abyss to abyss, from world to world, from heaven to heaven, from ecstasy to ecstasy. In that fire, all things were revealed, all mysteries were made clear."[73]

From a scientific viewpoint, Dr Walach suggests "that radical introspection or spiritual experience is also a way to discover the deep moral structure of reality and moral absolutes, if there are any. (…) The essence of such experience seems to be that all beings are interconnected in a unity of being and as such what we do to others we do to ourselves, and the other way round. This is the experiential basis for the traditional ethical statement: 'Love your neighbour as yourself.' "[74]

In terms of the wisdom teaching, as Dr Mac was told: "Understand the impersonal unity in all things, disregard personal separation, live in the conscious realisation of your oneness with the Creator of all mankind. Love your neighbour as yourself."[75] And later: "As long as relationship is not understood there can be no right action; there can be no solving of the problems which is of the self. There can be no relationship without self-knowledge. Only with knowledge of the self is there wisdom, and with wisdom there is Love."[76] Or, in the teaching conveyed by a Venusian Brother to Finnish ballerina and contactee Margit Mustapa: "All that is beyond the mind is love!"[77] (To be sure, the Teachers are not referring here to our sentimental, romantic notion of love, which is really a poor reflection of the universal energy that holds everything together.) Likewise, French contactee Pierre Monnet was told: "Love can only be lived! Love becomes visible through its radiation. Love forgets itself, gives itself to the other completely. Loving is easy and difficult at the same time. It is easy if one is prepared to look deep inside oneself and stare oneself straight in the face. Then one takes oneself by the hand and pulls oneself from the morass of actions and thoughts that strangle us and do not allow us to breathe freely and live in love. But it is difficult if we do not want to make the effort. "[78]

Enrique Barrios, a Chilean student of the wisdom teaching who had a profound experience of contact with a visitor from space in 1985, was taught: "Love is a beneficent ingredient of consciousness. It is capable of showing the most profound sense of existence. Love is the only legal 'drug'. Some mistakenly search in liquor and other drugs that which is produced only by Love. (…) There is no need to call for it, because it is already within us. You must not ask it to arrive, instead you must let it go, liberate it, give it freely."[79]

With regard to inducing experiences of oneness or expansion of consciousness artificially, others too have warned against the use of hallucinogens – the Masters, for instance, do not see the use of recreational drugs, including marijuana, as harmless. In the words of Benjamin Creme, "Drugs are very much contra-indicative to any kind of spiritual path because they ruin the nervous system."[80] And Adamski warned: "Many mystics have employed stimulants of various kinds to promote the knowledge of the unknown. But this is temporary, and many times only a hallucination or a reflection of their desires. This of course is unnatural and does not lead to beneficial knowledge of oneself. And only by knowing and following the natural laws [of life and evolution] can lasting knowledge be obtained. That is why the words of wisdom, 'man know thyself and you shall know all things,' have come down through the ages, and are just as true today as when they were spoken. And if we live these lessons, instead of stumbling in the dark as so many have done, we can go direct to the library of the Cosmos, the house of knowledge, directed by consciousness."[81]

In their attempts to describe the experiences of oneness that touched the depth of their souls the contactees relied on words and phrases that will always seem religious, new age-y, or mystical to 'nuts-and-bolts' UFO researchers, who classed them as a new kind of 'UFO religion' and dismissed them along with the numerous channelled messages from fanciful 'galactic commanders' that popped up in the slipstream of the contactees. But what is experienced here could well be what Dr Walach refers to as spiritual experiences that 'men are not allowed to speak of': "This we take to mean that the content of such experiences is too rich to be pressed in the structure of simple sentences, and to be submitted to exclusive logic. (…) This may be due to the fact that such direct, radically introspective experiences of reality touch upon the deep structure of reality from within." He then quotes from the experience that well-known astrophysicist Alan Lightman, a self-declared atheist, had while travelling to his island residence in Maine by boat: "I lay down in the boat and looked up. A very dark night sky seen from the ocean is a mystical experience. After a few minutes, my world had dissolved into that star-littered sky. The boat disappeared. My body disappeared. And I found myself falling into infinity. (…) I felt an overwhelming connection to the stars, as if I were part of them. And the vast expanse of time … seemed compressed to a dot. I felt connected not only to the stars but to all of nature, and to the entire cosmos. I felt a merging with something far larger than myself, a grand and eternal unity, a hint of something absolute." [82]

Thus, in the context of the broader view of reality presented in these pages, it should be clear that with the right intention and a clear focus it is not only necessary, but also possible to distinguish between authentic experiences and figments of an overactive imagination.

Experiences such as these have such a profound effect because they stir not only the soul, but even the core of every cell in our being through the interconnectedness from the Source of Being to the densest physical manifestation of Becoming. An explanation from Mme Blavatsky's Teacher may serve to illustrate this: "The one element not only fills space and is space, but interpenetrates every atom of cosmic matter." [83] And so, "How then can we doubt that a mineral contains in it a spark of the *One* as everything else in this objective nature does?" [84]

Here we are reminded of the spiritual force that Greek philosopher Plotinus called "The One", as referenced by Steve Taylor in his book *Spiritual Science*: "Each being contains in itself the whole intelligible world. Therefore All is everywhere... Man as he is now has ceased to be the All. But when he ceases to be an individual, he raises himself again and penetrates the whole world."

As an interdisciplinary approach to the study of systems, from simple to complex, systems science, in the words of professor Ervin Laszlo, finds it is precisely towards this innate oneness that evolution seems to be aiming: "The evident purpose of evolution in [the universe] is to achieve coherence in the domain of natural systems, and embracing oneness in the sphere of consciousness."[85]

The same direction can be found when looking at the evolution of the physical form. As Laszlo says: "Consciousness is pulled from the future rather than pushed by the past." In other words, the underlying oneness of reality is pulling us, our conscious awareness forward to ever greater expression of itself: "Through the action of love, an evolved consciousness fuses the elements of the mind into a higher unity. Attaining this 'spiritual evolution' is the meaning of existence."[86]

When we compare this scientific approach with the findings of the Ageless Wisdom teaching, we find that already in 1932 George Adamski taught something very similar: "...we are merely improving on our instruments, our bodies, making them more and more sensitive, so that, in time, through constant improvement, we can contact Divine Mind..."[87] Before him, one of Mme Blavatsky's Teachers wrote in 1882: "The evolution of the worlds cannot be considered apart from the evolution of everything created or having been on these worlds."[88] Here, too, the experience and teaching of the visitors from space agree, when George Adamski quotes a Master from Venus whom he met: "Understanding of the universal law both uplifts and restricts. As it is now with us, so it could be on your Earth. Lifted up by your knowledge, this same understanding would make it impossible for you to move in violence against your brothers."[89]

Similarly, her extraterrestrial teacher taught Margit Mustapa: "Present civilization will understand the uplifting love quality through group love, calling for group activities and unity in thinking. It does not mean *uniform* thinking or a certain pattern, or way of living. Brotherhood is composed of different individuals, expressing their own personal realizations of a new dimension of thought. The living substance of unity is as old as life itself in brotherhood"[90] – i.e. with respect for each individual's experience, talent and destiny.

Or, as Adamski noted in *Pioneers of Space*: "It was brought out that no where in the vast universe that they have thus far penetrated have they found an exact duplication. There is much similarity but infinite variety. It seems that the whole thing sums up into one, and that is like a vast university of endless numbers of grades."[91] And elsewhere: "On our planet, and on other planets within our system, the form which you call 'Man' has grown and advanced intellectually and socially through various stages of development to a point which is inconceivable to the people of your Earth. This development has been accomplished only by adhering to what you would term the laws of Nature. In our worlds it is known as growth through following the laws of the All Supreme Intelligence which governs all time and space."[92]

The same sense of interdependent individuality we find in the account of Adamski's Brazilian fellow contactee Dino Kraspedon: "Our Universe (…) is itself only an island within the Infinite, perhaps little more than one of the grains of sand that desert winds carry to far off places; we do not know where the winds come from, nor whither they are going. (…) To an amoeba, a drop of water must appear infinite, and it could not even conceive of the Earth that sustained it. In a sense the amoeba would be right, as the drop of water marks the limits of its consciousness, but not the limit of life. In relation to the Infinite, what more are we and our little world than the amoeba in its drop of water."[93]

Seen thus, unity in diversity is the only viable way for the manifestation of our innate oneness as a species, and of our oneness with Earth as the living organism of which we are a part. When this diversity is ignored, corrupted or destroyed, this will inevitably have its effects within the totality of the organism, as recent developments have shown: climate crises, the outbreak of the Covid-19 pandemic and other highly infectious diseases, and the terrible consequences of rising global inequality. A recent article by the director general of WWF International, the executive secretary of the UN Convention on Biological Diversity, and the director of the World Health Organization department of environment, climate change and health, states unequivocally that pandemics such as coronavirus are the result of humanity's destruction of nature. Their report leaves no doubt about the connection between our lack of regard for the natural order of diversity and its disastrous impact on our immune system: "Nature is currently declining globally at rates unprecedented in human history, and this is actually increasing our vulnerability to new diseases, particularly as a result of land-use change through activities such as deforestation, and agricultural and livestock intensification. These outbreaks of disease are manifestations of our dangerously unbalanced relationship with nature."[94]

With philosopher Charles Eisenstein we might wonder: "What is it that makes the vast majority of humanity comply with a system that drives Earth and humankind to ruin? What power has us in its grip? It isn't just the conspiracy theorists who are captive to a mythology. Society at large is too. I call it the mythology of Separation: me separate from you, matter separate from spirit, human separate from nature. It holds us as discrete and separate selves in an objective universe of force and mass, atoms and void. Because we are (in this myth) separate from other people and from nature, we must dominate our competitors and master nature. (…) That same myth motivates the conquest and ruin of nature, organizing society to turn the entire planet into money – no conspiracy necessary."[95]

In a time of pandemic the result of this 'myth of separation' led to rising hunger and poverty among the poorest in many European countries, including Greece, Italy, but also in the UK. In the US, the Center on Budget and Policy Priorities reported [July 2020] that the loss of employment income, arrears in mortgage payments, and rising food costs "have driven shocking increases in food insecurity affecting children, with over one-quarter of households with children reporting food insecurity in recent weeks."[96] Worldwide, the UN's World Food Program (WFP) warned of famines of "biblical proportions" as the Covid-19 pandemic would likely push an extra 130 million people into near-starvation. In May 2020 the WFP reported it was "providing food to nearly 100 million people every single day – and of that number, around 30 million depend on them for their very survival". According to their analysis, "300,000 people could starve to death each day in the coming months, if their ability to provide this normal support is disrupted. That's not including those who are newly impoverished by the pandemic itself."[97]

This stands in stark contrast with the manifestation of

brotherhood as it is expressed in the societies of the visitors from space, that I documented in my book *Priorities for a Planet in Transition*. In Chapter 3, 'Right human relations – An extraterrestrial show-and-tell', I compiled relevant observations and statements from various contactees, which all bear a striking similarity in their descriptions of how the oneness of life needs to be respected and actively expressed in the way society is organised. As George Adamski noted, humanity assumes that "the planet belongs to Man, with each claiming his small plot; while our neighbors in space realize their planet belongs to the Creator. So as one large family they share its products equally."[98]

In his eloquent essay Charles Eisenstein indicates what inhibits a similar approach on Earth: "The virus we face here is fear, whether it is fear of Covid-19, or fear of the totalitarian response to it, and this virus too has its terrain. (…) This terrain can be changed (…) by systemic change toward a more compassionate society, and by transforming the basic narrative of separation: the separate self in a world of other, me separate from you, humanity separate from nature."[99]

Readers may be familiar with William Golding's 1954 novel *Lord of the Flies*, which depicts the dark depths of human selfishness and cruelty when a group of British schoolboys find themselves marooned on an uninhabited island in the Pacific Ocean. Despite their best intentions, the boys' resolve to make the best of their situation until they are found soon disintegrates into savagery. However, when six boys from Tonga were shipwrecked for 15 months *in real life* in 1965 an entirely different story unfolded.

Dutch journalist Rutger Bregman, who rose to world fame when he spoke the unadulterated truth to the annual meeting of the powerful in Davos in January 2019 (look for 'Rutger Bregman Davos' on YouTube), researched the story about the Tongan boys and concluded that "It's time we told a different story... The real *Lord of the Flies* is a tale of friendship and loyalty; one that illustrates how much stronger we are if we can lean on each other."[100]

Scientific evidence for the innate oneness of humanity – albeit circumstantial – may be found in a growing body of research which shows that competition or war, contrary to popular belief, may not be "just human nature" after all, but rather a result of conditioning. In 2013 two anthropologists of the Abo Akademi University in Vasa, Finland, found evidence challenging the idea that war is the result of an evolutionary ancient predisposition that humans inherited. Studying how humans lived for more than 99.9 per cent of human history among 21 mobile bands of hunter-gatherer societies, they found that war is actually "an alien concept" and "only a tiny minority of violent deaths come close to being defined as acts of war".[101]

Although their findings were contested by other researchers, based on a further study in 2018 professor Brian Ferguson of Rutgers University Newark, too, argued in *Scientific American* "that war may not be in our nature at all". He concludes that "it is the overall circumstances that we live in that creates the impulse to go or not to go to war".[102]

In fact, in his book *No Contest. The Case Against Competition*, educationalist Alfie Kohn devoted an entire chapter to debunking the myth that even competition is an unavoidable fact of human life and part of "human nature". He says: "Arguably, the ubiquity of cooperative interactions even in a relatively competitive society is powerful evidence against the generalization that humans are naturally competitive."[103]

His observation is supported by Steve Taylor who counters the materialist argument that humans are ruthless genetic propagation machines: "...if living beings are made up of selfish genes, why do they often behave so unselfishly?" Based on findings from various studies, he concludes that altruism and co-operation are much more innate than competition and aggression.[104]

As an example of how this knowledge may be used to our advantage and encourage us to demonstrate our innate oneness in terms of practical living, the space visitors told Adamski that on their planets, "All production is for the benefit of everyone, with each receiving according to his needs. And since no medium of exchange such as money is involved, there are no 'rich'; there are no 'poor'. But all share equally, working for the common good. This may well be called a system of production for use."[105]

Some would dismiss the idea that this could ever be applied on Earth as idle hope, pointing at the polarization of societies everywhere. However, sceptics should ask themselves if the current assault on human lives as disposable production units for the greater bank balance of multibillionaires and the fragile egos that are drunk with the power awarded them by disenfranchised people living in defective democratic systems, would be so fierce and unrelenting if they didn't fear humanity's growing response to our oneness? Even though it is largely unreported, and what is reported appears to exhibit increased animosity and divisions, more and more people are waking up to the underlying oneness of life and uniting to bring their new awareness into manifestation, as we will see in the next chapter.

The original Oneness of life needs to be manifested as unity which, as we saw, is nothing to do with uniformity but with the richest possible diversity for any demonstration of unity to be meaningful. Diversity, says Benjamin Creme, "is the fundamental nature of the life of humanity. The individuality of every human being is not only a fact; it is one of the great facts of human evolution. Individuality shows the uniqueness of every person. (...) The greatest diversity within the greatest unity, or put the other way, the greatest unity with the greatest diversity, is the ideal that humanity is seeking, and is in alignment with the Plan of our Logos for the development of this world."[106]

Unity in diversity is the only valid and viable way to give expression to the underlying reality and beauty of life: Oneness. Ultimately, our experience of life is a reflection of what we identify with. Do we identify with our limited, physical self, with the material notion of success, the separatist notion of our ethnicity or nationality? Or do we identify with the soul, the seat of consciousness that connects us with the source of Existence, with the betterment of our community or society, with the human race and its humanity?

On his way home from the moon in 1971, Apollo 14 astronaut Edgar Mitchell had a spiritual experience when he saw the earth, the sun, the moon in a 360 degree panorama of the heavens. The magnificence of this view triggered a profound sense of ecstacy and identification with the whole: "My understanding of the distinct separateness and relative independence of movement of those cosmic bodies was shattered. I was overwhelmed with the sensation of physically and mentally extending out into the cosmos. The restraints and boundaries of flesh and bone fell away." This experience helped him realize "that the story of ourselves as told by science – our cosmology, our religion – was incomplete and likely flawed. I recognized that the Newtonian idea of separate, independent, discreet things in the universe wasn't a fully accurate description. What was needed was a new story of who we are and what we are capable of becoming."[107]

Through the ages, the way toward expanding our identification and experience of reality has been pointed by the pioneers of Oneness – the Masters of Wisdom, who are the Elder Brothers of humanity, and the Space Brothers. The former are those members of the human kingdom who have gone before us on the path of evolution and found the ways to reconnect with the Source and express it at a level that then becomes the next goalpost for humanity.

At this crucial time in the evolution of the planet and the human kingdom, the Space Brothers, often similarly advanced beings from other planets and pioneers in very much the same sense, are here to assist the Masters and humanity in greater numbers than ever, even though they take great care not to force their presence on the unprepared minds of their erratic younger brothers of Earth.

As this chapter shows, research and personal experiences lead an increasing number of scientists to conclude that consciousness is fundamental to life as we know it. This has been a central premise of the Ageless Wisdom teaching since it was reintroduced to the modern world.

Taken to its logical conclusion, the wisdom teachings' premise of a kingdom of nature that has evolved from the human kingdom should be accepted at least as a working hypothesis. As a student of the wisdom teaching for more than 40 years and based on my own experiences, to me the Masters of Wisdom are a reality. For these reasons, I feel justified in presenting Their information alongside the insights of science and the visitors from space.

References

1 'A Biographical Sketch of the life of Dr Murdo MacDonald-Bayne by Paul Troxler and Lora Mendel'. See: <www.thomasstaudtverlagflensburg.de/home/english/biography-by-paul-troxler-and-lora-mendel/>

2 Eben Alexander MD (2012), 'My experience in coma'. See: <ebenalexander.com/about/my-experience-in-coma/>

3 George Adamski (1964), *Science of Life* study course, Lesson Eleven, 'Exploration of Cosmic Space'

4 Adamski, 'Private Group Lecture for Advanced Thinkers'. Detroit, MI, USA, 4 March 1955, p.3

5 'Tibetan Monastery, First in America, to Shelter Cult Disciples at Laguna Beach'. *Los Angeles Times*, 8 April 1934

6 Adamski, letter to Emma Martinelli, 16 August 1950

7 A. Trevor Barker (comp.; 1923), *The Mahatma Letters to A.P. Sinnett*, p.90

8 Benjamin Creme (1997), *Maitreya's Mission*, Vol. Three, p.22

9 Michio Kaku, as quoted in Nina L. Diamond (2000), *Voices of Truth: Conversations with Scientists, Thinkers and Healers*

10 Ervin Laszlo (2016), *What is Reality? The New Map of Cosmos and Consciousness*, p.44

11 Paul Howard (dir., 2020), *Infinite Potential: The Life & Ideas of David Bohm*

12 H.P. Blavatsky (1888), *The Secret Doctrine*, Vol. I, p.43 (6th Adyar ed. 1972, Vol.1, p.115

13 Ibidem, p.17 (6th Adyar ed. 1972, Vol.1, p.82)

14 Ibid., p.70 (6th Adyar ed. 1972, Vol.1, pp.137-38)

15 Blavatsky (1889), *The Key to Theosophy*, pp.83-84

16 Klee Irwin, Marcelo Amaral and David Chester, 'The Self-Simulation

Hypothesis Interpretation of Quantum Mechanics', p.16. 12 February 2020. See: <www.mdpi.com/1099-4300/22/2/247/pdf>

17 Ibidem, p.22

18 Laszlo (2017), *The Intelligence of the Cosmos*, p.35

19 Laszlo (2016), op cit, pp.20-21

20 B. Kumar (1986), *Origin and Evolution of Mankind*, pp.23-24

21 Laszlo (2016), op cit p.19

22 Sylvia Cranston (1993), *HPB – The Extraordinary Life and Influence of Helena Blavatsky, Founder of the Modern Theosophical Movement*, p.xx and pp.557-58

23 Perimeter Institute for Theoretical Physics, 'Is the Universe a Bubble? – Physicists Work on the Multiverse Hypothesis', 21 July 2014. See: <scitechdaily.com/universe-bubble-physicists-work-multiverse-hypothesis/>

24 Laszlo (2016), op cit, p.9

25 Adamski (1964), op cit, Lesson Seven, 'Cosmic Memory'

26 Adamski (1949), *Pioneers of Space*, pp.5-6

27 Ibid., p.185

28 Ibid., pp.185-86

29 Benjamin Creme (2001), *The Great Approach – New Light and Life for Humanity*, p.129

30 Blavatsky (1889), op cit, pp.85-86

31 Wilbert Smith (1969), *The Boys from Topside*, p.28

32 Howard Menger (1959), *From Outer Space to You*, p.172

33 Blavatsky (1889), op cit, pp.88-89

34 Dino Kraspedon (1959), *My Contact With Flying Saucers*, p.161

35 Barker (1923), op cit, p.66ff

36 Menger (1959), op cit, p.175

37 Adamski, letter to Emma Martinelli, 24 November 1951

38 Steve Taylor (2018), *Spiritual Science. Why science needs spirituality to make sense of the world*, p.52

39 Kumar (1986), op cit, p.36

40 Blavatsky (1889), op cit, p.87

41 Murdo MacDonald-Bayne [n.d.; 1954], *Beyond the Himalayas*, p.161

42 Creme (2002), *The Art of Co-operation*, pp.115-16

43 Vera Stanley Alder, (1939), *The Initiation of the World*, p.36

44 MacDonald-Bayne [n.d., 1954]), op cit, p.161

45 Sarah Durston and Ton Baggerman (2017), *The Universe, Life and Everything*, p.9. See: <library.oapen.org/handle/20.500.12657/31132>

46 *The X Factor* (1989), 'Carlos Díaz'. Issue 89, p.2469

47 Interview with Carlos Díaz in Michael Hesemann (dir.; 2001), *Ships of Light – The Carlos Díaz Experience*. Available at <www.youtube.com/watch?v=SaAluEBxXLI>

48 Interview with Pierre Monnet, *L'Ère Nouvelle*, January 2006. See: <pointdereference.free.fr/m/www.erenouvelle.com/PORTCO-7.HTM>. Author's translation from French.

49 Benjamin Creme's Master (2011), 'Man's responsibility'. In: Creme (ed.; 2017), *A Master Speaks*, Vol. Two, pp.157-58

50 Sonia Shah, 'Think Exotic Animals Are to Blame for the Coronavirus?

Think Again', 18 February 2020. See: <www.thenation.com/article/environment/coronavirus-habitat-loss/>

51 Phoebe Weston, '"We did it to ourselves": scientist says the intrusion into nature led to pandemic', 25 April 2020. See: <www.theguardian.com/world/2020/apr/25/ourselves-scientist-says-human-intrusion-nature-pandemic-aoe>

52 Dawna Jones, 'What is reality? Interview with Dr Ervin Laslzo', 28 November 2016. See <www.youtube.com/watch?v=1Ke2apZ5aPk>

53 Richard Barrett (2018), *The Evolutionary Human. How Darwin Got It Wrong*, p.19

54 Taylor (2018), op cit, p.183

55 David Loye (2004), *Telling the New Story*, pp.5-7. Available at <www.thedarwinproject.com/adventure/newstory/newstory.pdf>

56 Charles Darwin (1871), *The Descent of Man*, as quoted in David Loye (2018), *Rediscovering Darwin. The Rest of Darwin's Theory and Why We Need It Today*, p.106

57 Creme's Master (2008), 'Evolution versus creationism'. In: Creme (ed.; 2017), *A Master Speaks*, Vol. Two, pp.111-12

58 Adamski (1964), op cit, Lesson Five, 'Consciousness, The Intelligence And Power Of All Life'

59 Adamski, letter to Emma Martinelli, 13 March 1950

60 Interview with Pierre Monnet, op cit

61 Alder (1979), *From the Mundane to the Magnificent – A Volume of Autobiography*, pp.69-70, 72

62 Evelyn Fox Keller (1983), *A Feeling for the Organism. The Life and Work of Barbara McClintock*, p.117

63 Adamski (1964), op cit, Lesson Four, 'The Relationship of All Creation'

64 Adamski (1949), op cit, pp.127-28

65 Ibid., p.63

66 Ibid., p.118

67 Ibid., pp.156-57

68 MacDonald-Bayne [n.d.; 1954], op cit, p.21

69 Adamski (1949), op cit, p.157

70 Ibid., p.211

71 Adamski (1962), *My Trip to the Twelve Counsellors Meeting That Took Place on Saturn,* Part 1, pp.3-4

72 Ibid., Part 2, p.1

73 Giorgio Dibitonto (1990), *Angels in Starships*, pp.107-08

74 Harald Walach (2019), *Beyond a Materialist World View. Towards an Expanded Science*, p.83

75 MacDonald-Bayne [n.d.; 1954], op cit, p.77

76 MacDonald-Bayne [n.d.; 1956], *The Yoga of the Christ*, p.138

77 Margit Mustapa (1963), *Book of Brothers*, p.119

78 Monnet (1994), *Contacts d'Outre Espace.* Author's translation from the Dutch edition (1995), *Een boodschap van vrede*, pp.130-31

79 Enrique Barrios (1987), *Ami Returns*, Chapter 14

80 Creme (1985), *Transmission – A Meditation for the New Age*, 4th ed. 1998, pp.82-83

81 Adamski (1964), op cit, Lesson Eight, 'Cosmic Oneness'

82 Walach (2019), op cit, pp.80-81

83 Barker (1923), op cit, p.97

84 Ibid., p.93

85 Laszlo (2017), op cit, p.46

86 Ibid., p.41

87 Adamski (1932), *The Invisible Ocean*, p.15

88 Barker (1923), op cit, pp.72-73

89 Adamski (1955), *Inside the Space Ships*, p.93

90 Mustapa (1963), op cit, p.154

91 Adamski (1949), op cit, p.228

92 Adamski (1955), op cit, p.86

93 Dino Kraspedon (1959), op cit, p.161

94 Marco Lambertini, Elizabeth Maruma Mrema, and Maria Neira, 'Coronovirus is warning us to mend our broken relationships with nature'. *The Guardian*, 17 June 2020. See: <www.theguardian.com/commentisfree/2020/jun/17/coronavirus-warning-broken-relationship-nature>

95 Charles Eisenstein (2020), 'The Conspiracy Myth'. See: <charleseisenstein.org/essays/the-conspiracy-myth/>

96 Brynne-Keith Jennings, 'Number of Food-Insecure Households With Children Doubled From Pre-COVID Levels. CBPP, 8 July 2020. See: <www.cbpp.org/blog/boosting-snap-needed-to-reduce-hardship-long-term-effects-on-children>

97 Zaria Gorvett, 'Why most Covid-19 deaths won't be from the virus'. *BBC Future*, 28 May 2020. See: <www.bbc.com/future/article/20200528-why-most-covid-19-deaths-wont-be-from-the-virus>

98 Adamski (1957-58), *Cosmic Science*, Part No.1, Q9

99 Eisenstein (2020), 'The Coronation', See: <charleseisenstein.org/essays/the-coronation/>

100 Rutger Bregman, 'The real Lord of the Flies: what happened when six boys were shipwrecked for 15 months'. *The Guardian*, 9 May 2020. See: <www.theguardian.com/books/2020/may/09/the-real-lord-of-the-flies-what-happened-when-six-boys-were-shipwrecked-for-15-months>

101 Steve Connor, 'Is it natural for humans to make war? New study of tribal societies reveals conflict is an alien concept'. *The Independent*, 18 July 2013. See: <www.independent.co.uk/news/science/is-it-natural-for-humans-to-make-war-new-study-of-tribal-societies-reveals-conflict-is-an-alien-8718069.html>

102 Rutgers University, 'No scientific proof that war is ingrained in human nature, according to study', 4 December 2018. See: <phys.org/news/2018-12-scientific-proof-war-ingrained-human.html>

103 Alfie Kohn (1992), *No Contest. The Case Against Competition*, pp.18-19

104 Taylor (2018), op cit, pp.192-199

105 Adamski (1957-58), op cit, Part No.1, Q18

106 Creme (2012), *Unity in Diversity. The Way Ahead for Humanity*, pp.38-39

107 Cassandra Vieten, 'In Memoriam: Edgar Mitchell 1930-2016'. IONS Blog, 5 February 2016. See: <noetic.org/blog/in-memoriam-edgar-mitchell-scd-phd/>

"You are never given a dream without also being given the power to make it true. You may have to work for it, however."
—Richard Bach, author

3. BACK TO REALITY:
THE ALGORITHMS OF EVOLUTION

Early July 2020 saw a mild euphoria in the media when in a matter of days "a flotilla of probes" built by the USA, China and the United Arab Emirates were sent on separate, seven-month voyages to explore Mars: "Never has so much interplanetary traffic been put en route to Mars at one time – and all of it is intended to help answer a question that has nagged scientists for decades: is there, or was there ever, life on Mars?"[1] Even if the scientists behind these efforts are sincere in their quest, we may be sure that the people who have signed off on their projects, or their financial backers, will be more interested in what can be sourced from the red planet than in finding signs of life. A news report about bad weather delaying the UAE's mission in fact hinted at the ulterior economic motive: "The country's oil reserves are not inexhaustible, as the United Arab Emirates authorities well know. That knowledge is one of the underlying motives for their massive innovation programmes."[2]

Nonetheless, as George Adamski learned, "Venusian scientists are hopeful that men of earth will settle down to a constructive program and succeed in building ships that can travel space. For then they can visit other planets and bring the earth into a proper relationship with them. (...) The cosmos is filled with wonders yet to be witnessed. Earth men have engulfed themselves with a bondage of self – and ego assertions – when they know practically nothing about the finer, inner workings of creation."[3]

Manifesting our dawning sense of Oneness, however, could hardly be about monopolizing business or colonizing the solar system (even only at the dense-physical plane) as some big tech 'visionaries' like Jeff Bezos and Elon Musk seem to think. For instance, Jeff Bezos has indicated that the reason behind his drive to make Amazon.com the largest company on which consumers rely for everything from daily groceries to television shows and films, with its own network of satellites to deliver it all, is his aim to create giant space colonies through his other company, Blue Origins. Mr Bezos seems to think that the only possible growth is material growth: "The earth is finite, and if the world economy and population is to keep expanding, space is the only way to go." He asks: "What happens when unlimited demand meets finite resources? Rationing." But, he says, "if we move out into the solar system, for all practical purposes, we have unlimited resources."[4]

So in Mr Bezos' philosophy of 'too much of everything is not enough' the solution to the problems we created by ignoring the spiritual reality that Life is One, and therefore we are one with planet Earth and its place among the stars, is to take the same approach of indiscriminate and unfettered exploitation and squandering, and replicate this mistake on a cosmic scale in the solar system. As *The Atlantic* puts it, "to say that Bezos's ultimate goal is dominion over the planet is to misunderstand him. His ambitions are not bound by the gravitational pull of the Earth."[5]

In their contacts with individuals on Earth the Space Brothers often cautioned us about transgressions against our oneness with Nature. For instance, Adamski's Venusian contact Orthon said: "...although the laws which govern the relationship of man to the world on which he lives would not at this present time be understood by men of Earth, I want to stress that the erring path which they have followed so consistently is actually the reason for their ignorance of your planet's present instability."[6] This was underscored by Benjamin Creme's Master who wrote: "Men must realize their responsibility for the planet on which they live. Stewards, men are, of a strong but sensitive organism and must protect it from harm. Few, today, can claim that this they do. On the contrary, men waste and ride roughshod over nature's generous munificence, unheedful of tomorrow or their children's needs."[7]

A Master from Saturn explained to Adamski: "Since we have learned that life is all-inclusive and that we *are* that life, we know that we can hurt nothing without hurting ourselves."[8] In practical terms this means, Adamski says, "They know that all life-forms are important in the Cosmic Plan, and without human interference Nature provides amply for all her children, yet maintains an eternal equilibrium."[9]

Underlying everything that exists is consciousness that bends waves into clusters of vibration suited to the specific purpose of each cluster. These clusters may be the atom in a body cell, an organ in the human body, a kingdom of nature in the life of a planet, a planet in a solar system, a solar system in a galaxy, and so on. And in the immeasurable vastness of the Laniakea supercluster that is home to our Milky Way, these are all still forms that would not be there without the underlying cause that informs them. Once they have fulfilled the purpose of their original cause, the form will disintegrate while that which caused it to manifest will still be there, but in a cycle of suspended activity.

In the previous chapter I referenced a recent paper hypothesising that "the physical universe, as a strange loop, is a mental self-simulation".[10] Still rooted in the prevailing 'materialist' approach to understanding the universe, the paper stops short of theorizing who or what the originating "self" might be. This begs the question where these forms originate. If they are a "self-simulation", or "self-originated", who or what is that Self? That, of course is the final question, similar to asking 'where did God come from?' or 'why did God create the Universe?'. It is the question to which there will be no answer until we get there, back to the Source. If ever there was a question whose answer was in the asking, it must be this. Another way of putting it is the Delphi Oracle's maxim "Know thyself", or the Indian avatar Ramana Maharshi's "Who am I?"

One way to approach this question is by first establishing that the very notion of 'self' implies something that is 'not-self', everything outside and around the discrete point of awareness in the sea of consciousness that, in the case of humans, we call "me". As Annie Besant, the women's rights activist who succeeded H.P. Blavatsky as President of the Theosophical Society, describes it, consciousness is "life becoming aware of its surroundings".[11]

In the process of creation and evolution – the Source (or God) externalizing its very essence into the farthest reaches of dense-physical objectivity – the human kingdom forms the stage where its focus, that has long been extended outward, inevitably reaches the 'event horizon' of objectivity and starts its journey back to its true Self, as an aspect of the Source. Then begins the search, now in growing self-awareness, back to its original state of Oneness, by seeking ever greater unity and synthesis.

The British author and student of the wisdom teachings Vera Stanley Alder describes it thus: "Physical life expresses itself by means of infinite divisions and separations, forming ever-increasing numbers of isolated objects, their isolation being caused by the fact that in the lower dimensions of form life, each object requires its own space, and there is no interpenetration." As a result, she adds, "Separatism finally gained utter sway, and man's earlier recognition of the unity of life faded completely from his memory. It was the era of self-development, of extreme egotism and intense sacrifice to the three dimensions [i.e. the physical state of independent motion]. (…) Unfortunately, however, the more man concentrated upon the third dimension, the more he became separated from his hitherto involuntary recognition of and awareness of the higher dimensions."[12]

Having reached its zenith, this process of division and separation can now be witnessed in an unprecedented number of people claiming individuality and celebrating idealism of one kind or another. In fact, for many people in Western nations, individuality *is* their ideal and the current selfie culture or Instagramification of life could be seen as the apotheosis of the process of separation among modern humanity – that necessary phase in evolution – misguidedly attempting to objectify and idolize the 'self', rather than going inward to seek out and return to the Self.

Philosopher and educator J. Krishnamurti expressed it in his usual direct and lucid manner: "To live is to be related. It is only in the mirror of relationship that I understand myself – which means that I must be extraordinarily alert in all my thoughts, feelings and actions in relationship. This is not a difficult process or a superhuman endeavour; and as with all rivers, while the source is hardly perceptible, the waters gather momentum as they move, as they deepen. In this mad and chaotic world, if you go into this process advisedly, with care, with patience, without condemning, you will see how it begins to gather momentum…"[13] Sorely lacking this understanding, is it any wonder that people are electing unfit wannabe leaders whose sole aim is to make their own ego great again at the cost of society and humanity?

The objectified ideal of individuality even causes some people to fear efforts towards international cooperation as a dictatorial 'New World Order'. They warn against the United Nations Organisation and its agencies as the first steps towards a world government, as if global problems like climate change, pandemics, the unbridled power of international corporations, and the endless suffering caused by unnecessary hunger, poverty and illness that are accepted as collateral damage of commercializing every aspect of life for the fallacy of perpetual economic growth, do not require a global solution that can only be arrived at and implemented through international cooperation.

In a lecture in September 1955 George Adamski already pointed out, however, that unity is the inevitable direction for humanity in a world that is more connected than ever: "There is something transpiring in this world that we never dreamed would take place in our time, but it is taking place... the tendency toward a United World." In fact, he said: "The United Nations has promise of that result..."[14]

Manifesting our oneness by achieving justice through the equitable redistribution of food, natural resources, technological know-how, et cetera, would necessarily involve the United Nations because, according to Benjamin Creme's Master, it is "the forum in which the voice of the smaller nations can be raised and heard. This is only possible when the Security Council, with its arbitrary veto, is abolished. (...) Then we will see the nations acting without restraints imposed by Great Power veto and financial inducement. Those who call loudest for democracy in foreign lands are strangely blind to its absence in the halls of the United Nations."[15]

Given the threat of fascism in the current political climate in various countries, let us emphasize again that oneness, or unity, or synthesis, should not be mistaken for uniformity. Based on his many experiences with visitors from space Howard Menger put it very simply: "The will of the Infinite Father is to EXPRESS in ALL dimensions, the love of the Infinite Father in ALL FORMS, COLORS, SOUNDS, TASTES, AND EXPRESSIONS."[16]

Likewise, a Venusian Master told Adamski that many paths lead upward, and illustrated how diversity is integral to the created universe: "Though one man may choose one [path] and a second man another, this need not divide them as brothers. Indeed, one may learn from the other, if he will. For in the vastness of the infinite creation, there is no one way that is the only way."[17]

In 2015, when I was working on *Priorities for a Planet in Transition*, economic and financial experts were expecting another, possibly final collapse of our current system because no structural changes were made to the debt-driven world economy after the crash of 2008 had been reverted by creating even greater debts. Five years on, the fall-out from the Covid-19 pandemic looks like it may be the unexpected trigger that will finally send the world economy in a tailspin.

Also, the corona crisis seems to engender at last the beginnings of a sense that we cannot act against nature without consequences – on a global scale not seen before. The climate crisis has, of course, been mounting for many decades and is of even greater gravity and scope, except that its consequences have not made themselves felt quite as urgently and equally around the globe yet. Foreign affairs commentator Simon Tisdall points out that the pandemic is not an event that is unconnected to other global problems. He thinks that the agenda for the post-pandemic world should be "revolutionary" and include "meaningful steps to address poverty and the north-south wealth gap, more urgent approaches to linked climate, energy, water and mass extinction crises and, for example, the adoption of so-called doughnut economics that measures prosperity by counting shared social, health and environmental benefits, not GDP growth."[18]

With the innate oneness of our planetary system

consistently violated over many decades and the consequences for the environment and the human immune system persistently ignored in not only business models and government policy, but also in our individual attitudes towards the natural systems that give and sustain our lives, the pandemic cannot be seen as anything other than the logical response to our individuality gone ballistic, and our sustained separatism from the larger organism that is our planet.

In an interview titled 'This is a catastrophe that has come from within', French sociologist and philosopher Bruno Latour sees the corona pandemic as an opportunity to rethink current attitudes that lie at the heart of our inertia to address global problems: "Covid has given us a model of contamination. It has shown how quickly something can become global just by going from one mouth to another. That's an incredible demonstration of network theory. (…) It shows that we must not think of the personal and the collective as two distinct levels. The big climate questions can make individuals feel small and impotent. But the virus gives us a lesson. If you spread from one mouth to another, you can viralise the world very fast. That knowledge can re-empower us."[19]

Nearly 70 years earlier, the Venusian Master with whom George Adamski had an audience while on a mothership said the global catastrophe of the world wars had created fear and confusion in the minds of humanity who hungered for knowledge of a way to deliver them. He, similarly, said: "I think the peoples of Earth would be amazed to find how swiftly change could come throughout the planet. Now that you have the medium for world-wide broadcasting, messages urging love and tolerance for all, instead of suspicion and censure, would find receptive hearts."[20]

Throughout history people have managed, sometimes after long and bloody struggles, to oust political or economic regimes that thrived on cruelty, usury, slavery, or other forms of oppression. Our current economic system, that is sustained merely by the grace of unprecedented levels of inequality and social injustice, is kept in place by a political system that we call democratic but is nothing of the sort. Massive amounts of money are spent on influencing populations to vote for politicians who design legislation that goes directly against the interests of the common good and the common people – the 99% – to protect and advance the interest of those who are in economic power – the 1%. Even in 1949 the Master Djwhal Khul (D.K.), who gave the bulk of the *Secret Doctrine* through Mme Blavatsky, wrote: "True Democracy is as yet unknown; it awaits the time when an educated and enlightened public opinion will bring it to power; towards that spiritual event, mankind is hastening. The battle of Democracy will be fought out in the United States. There the people at present vote and organise their government on a personality basis and not from any spiritual or intelligent conviction."[21]

The oneness of humanity as a planetary species has been strenuously ravaged by market forces that were unleashed on society in the 1980s, in response to mankind's mounting call for justice and freedom. The great social movements that have sprung up since the 19th century demanded labour rights, women's rights, children's rights, voting rights, citizens' rights, et cetera, manifesting our gradual growth in consciousness, whereby individual interests were transcended and seen to coincide with the interests of many others in the same position. Thus, the individual's awareness of their needs was expanded to include the needs of others.

This social development began to demonstrate clearly in the 19th century in response to the spiritual energies that accompany the externalization of the spiritual kingdom in nature – the initiates and Masters of Wisdom – at the beginning of a new cosmic cycle, because in evolutionary terms humanity is ready to take the next step in the evolution of consciousness. Of humanity's role in the scheme of evolution Benjamin Creme's Master gave this evocative description: "Man is a crucible in which is being created a new Being. In the fiery heat of experience man is gradually learning the ways of God. Slow and painful may be the early steps but in time the pace will quicken. Revelation after revelation will expand his consciousness, leading to a crescendo of creativity and knowledge. Man will stand revealed as a Son of God."[22]

Manifesting first in large groupings of people who organised to achieve 'strength in numbers', so as to make the 'voice of the people' heard loud and clear to lawmakers and legislators, this expansion of consciousness eventually precipitated into laws protecting the rights of citizens in their respective countries. This unifying voice found a resounding crescendo 70 years ago, in the ratification by the international community of the Universal Declaration of Human Rights (UDHR) in December 1948, extending basic rights to everyone simply by virtue of having been born into the human kingdom (Appendix III).

Politicians and the media like to parade their commitment to human rights when it suits their own political likes or dislikes of regimes that they support or oppose, and as a result the human rights we are used to seeing 'defended' relate mostly to protection against persecution or oppression based on "race, colour, sex, language, religion, political or other opinion".

Yet, of the 30 articles outlining every human being's rights, the one that explicitly extends protection against economic oppression – Article 25 – is never invoked by ruling or opposition parties when it comes to the plight of refugees or migrants seeking a liveable income and future prospects for themselves and their families. Often, their countries have been left at the mercy of corrupt local political systems, unfair international trade agreements, and neocolonial practices of global corporations. So, whenever armed conflict erupts in the Middle East, Africa or somewhere else – usually as a result of invisible or unnamed economic interests – the first thing Western politicians will do is promise refuge to "real" refugees and defence of their country's borders against "economic" refugees.

Dino Kraspedon's extraterrestrial teacher could have been speaking of this injustice when he was saddened "to see that wars take place against the wishes of most people, because poor people do not fight easily. Carnage has become the perquisite of the rich and powerful, of those who need no help and who even renounce God, seeing no necessity for the Divine Presence in their lives. Abundance blinded them, gluttony clouded their vision. Strife is the product of egotism. It cannot be said that they fight for principles, for a man of principle does not fight.

The great principles that have guided the life of many men on Earth, and which also guide life on other worlds, are love of God and of one's neighbour."[23]

Forty years ago, in 1980, the commission chaired by former German chancellor Willy Brandt presented the international community with a consensus solution for international cooperation to end poverty and inequality worldwide. Its report, *North-South – A Program for Survival*, acknowledged the injustices built into the post-war institutions and post-colonial trade agreements, which all favoured the ruling financial and economic powers and interests. It urged the world community to close the growing wealth gap between the developed and the developing part of the global community to relieve the international tensions that result from such prolonged and growing inequality.[24]

A year after the Brandt Report was published, at the economic summit of world leaders in Cancún, Mexico, Ronald Reagan countered Brandt's call for reform of the international institutions with his promise of getting government out of the way of private business interests after winning the presidency on a platform of 'opportunity for the individual'. As a result, regulations protecting consumers and laws protecting citizens and workers were curtailed or revoked, giving corporations free reign to squeeze every last penny of profit out of every basic human need, by endangering everyone's health through the relentless promotion of competition and consumption, and plundering the planet's natural resources. This has now reached the point where at last more and more people are waking up to the fact that this 'freedom', which the commitment to material pursuits at all costs was supposed to bring, is an illusion. And more and more people are beginning to see the dangers that result from the pervasive need for competition to survive in a world where obscene wealth only provides freedom for the 1 percent – freedom, that is, from the law, public scrutiny, due diligence, and accountability.

It is important in this respect to acknowledge we are not puppets on the strings of 'the Illuminati', the power brokers of a 'reptilian bloodline', or other sinister and all-powerful forces of darkness who some suspect to be in control of humanity. Notwithstanding their alleged thousands of years rule behind the scenes, they can't seem to get us into complete submission. If we, humanity, are not where we want to be or where we want our society to be, instead of seeking a scapegoat or saviour of alien or terrestrial origin, we should ask ourselves if we are doing what we can to change the situation? As philosopher Charles Eisenstein pointedly asks: "Of course, there are many bad actors in our world, remorseless people committing heinous acts. But have they created the system and the mythology of Separation, or do they merely take advantage of it?"[25]

In keeping with the neoliberal myth of 'individual freedom', Credit Suisse's Global Wealth Report touts wealth as "vital for economy and individuals", saying it allows citizens to "strive for more abstract goals and self-fulfilment, for example entrepreneurial activities". And while it admits this may be harder in countries with lower wealth levels, it cites the US as a positive example based on the number of new millionaires in 2018: "A good indication of whether a country offers a favorable environment for entrepreneurship is the number of millionaires in its population. The US, for example, tops the chart with the creation of 675,000 millionaires last year – more than half of the global increase."

What is left out of the Credit Suisse view of wealth as a driver of self-fulfilment is that the system in which the creation of wealth is essential for a decent life does not provide a level playing field for all its participants. The vast majority of people will never be able to reach the status of millionaire, or the economic freedom and security it provides, and are left to compete for a living against the odds. In fact, the zero-hour contracts that have been promoted as facilitating the "entrepreneurial spirit" of people, were meant mostly to limit workers' rights and maximize employers' and shareholders' profits. Besides, a system that sees competition as a healthy incentive for people to get ahead in life but allows over 40 million of its 'players' around the world to live in slavery or bondage[26], is really a display of the grossest degree of separation, not to mention inhumanity.

As outlined above, hard-won rights and protections have been scrapped and abandoned since the 1980s and as a result we now live in a world where the world's "10 richest billionaires (...) own $801 billion in combined wealth, more than the total annual production in goods and services of most nations, according to the International Monetary Fund." Since 1969 the income share of the top 1 percent has doubled, while the number of families living in poverty has remained more or less the same.[27] In stark contrast, a visitor from space told Daniel Fry in 1950: "With the lessons of the past constantly before our people, we have found it wise to always maintain the material values in proper relationship with the more important social and spiritual values."[28]

Perhaps it is for this reason that the second of the Delphic maxims guiding us to self-knowledge reads: "Nothing in excess." However, a poignant example from the Inequality.org website makes clear that even inequality is not distributed equally, leading to racial and gender inequality. For instance, as median wealth in the USA decreased slightly between 1983 and 2016, median wealth among whites increased by $36,000, while African Americans saw their median wealth *decrease* by nearly $2,000 over the same period. Overall, black and latino families in the US are twice as likely to have zero wealth.

Long before the neoliberal assault on the human right to social and economic security, after his exchange with a visitor from space in 1952, Dino Kraspedon saw the harsh truth about the way we have organised society on our planet: "If we do not, strictly speaking, eat each other's flesh, we live on the sweat of the poor in a disgraceful and unjust society, which progress will one day have to supplant; unless we kill ourselves off in a hydrogen war before that happens."[29]

When we accept a political system that enables and supports economic structures which depend on the violation of human dignity and oneness, indeed our inner spirit, it is no surprise that there is a noticeable increase in mental illnesses, depression and suicides. Because if you have no prospect of self-fulfilment, or even of feeding your family, or giving your children a future, to achieve some kind of economic security, to pay off your student loans or your credit card debts, or afford a decent health insurance, or pay your mortgage, the global increase in "wealth" or GDP is meaningless and merely destroys our environment. If you live in South or Central America, in Africa or parts of Asia, and you see no way out, you may decide to leave family, friends and what measly possessions you have that you call 'home', to at least have a chance of finding illegal and often dangerous work in the US, Canada, Britain, Europe, or Australia. That is, if you survive the trek through the desert or the sea journey with unscrupulous traffickers in an overcrowded boat, do not end up in the hands of modern slave traders, or in the sex industry, or die in a refrigerated truck. And if you live in destitution in the US, Britain, Europe, Australia or Japan and you see no way out, where do you go?

In the UK the introduction in 2013 of a new and simplified – and harsher – social benefits system, called Universal Credit, has been associated, according to the prestigious medical journal *The Lancet*, "with a 7 percent increase in psychological distress among recipients since the benefit was introduced – equivalent to an estimated 63,674 unemployed people. Of these, over a third – or 21,760 individuals – may have become clinically depressed, according to the researchers from the University of Liverpool."[30] As a result, *The Guardian* reported, at least 69 (!) suicides were linked to the way the Department for Work and Pensions handled benefits claims.[31]

Illegal drugs use and crime have long been the go-to escapes for people who don't see a way out of their misery, and both come at huge costs, not only for the individuals and families involved, but also for the rest of society. More recently, pharmaceutics abuse has become a less noticeable but equally destructive option to deal with depression, deliberately driven by Big Pharma's profit targets.

In the US, legislation usually only stands a chance of being passed if it includes provisions for privatization of public services and tasks, regardless of the political party in government, because that is the only way to secure a majority in either the House of Representatives or the Senate to pass a new bill. As a result, the situation in the US for those at the economic bottom is no better than in the UK or many other industrialized countries. According to *Vox*, if the corona crisis hasn't made matters worse, it "has revealed many uncomfortable truths about America, including the country's unemployment system: It is broken, and in many cases, it is broken by design. After years of disinvestment and underfunding, benefits systems across the country have been left starved and in disrepair. In many states, benefits are intentionally difficult to collect and application processes complex to navigate."[32] No wonder the people have lost their trust in government and become susceptible to conspiracies like the 'deep state' as a malevolent force behind the scenes of the democratic process.

The West's long-standing neglect of humanity's own brothers and sisters in developing countries is of course a gross violation of our innate oneness that has caused millions and millions of unnecessary deaths since colonies achieved political independence, but remained at the mercy of strict financial and economic dictates from the international trade system that was set up by the Western powers, such as the World Bank, the IMF, GATT/WTO and other major trade agreements. Global corporate powers and interests now largely dictate these, since being handed near-unlimited freedom in the 1980s celebration of 'free markets'. The rights that millions of people have struggled to secure for over a century now stand in the way of maximizing profits, and are relentlessly targeted by those who risk losing their hold on humanity when we realize our oneness and liberate ourselves from the need to compete for a living. No surprise, then, that his hosts on Mars told Adamski that our neighbours in space consider us as savage people: "We live off the miseries of each other."[33]

In the wake of large numbers of self-employed people on zero-hour contracts losing their insecure sources of income, there have been calls for the introduction of a universal basic income (UBI) in many countries. But in his recent book *Towards a universal basic income for all humanity*, Mohammed Mesbahi rightfully notes that "it is immoral for a full UBI to be implemented by one country alone (even if they have the means to do so), in a world where poverty is rampant and many war-ravaged nations are struggling to avert soaring levels of food insecurity." Mindful not just of the oneness of humanity but also of the inevitable mass migration that would follow, he asks: "It may well be possible for everyone in your own country to live comfortably with a guaranteed basic income and universal social services, but what about the others in poor countries who do not have such entitlements, or even a morsel to eat?"[34] Meanwhile, the United Nations Development Program called for the developing world's 2020 foreign debt payments to be diverted into a temporary basic income for the poorest 2.7 billion people. This would help slow down the surge in Covid-19 cases by guaranteeing the poorest food and shelter during the pandemic, but this appeal to our humanity seems to have been entirely ignored.[35]

Not for nothing, Bruna Latour suggests: "What we need is not only to modify the system of production but to get out of it altogether. We should remember that this idea of framing everything in terms of the economy is a new thing in human history. The pandemic has shown us the economy is a very narrow and limited way of organising life and deciding who is important and who is not important. If I could change one thing, it would be to get out of the system of production and instead build a political ecology."[36]

In *Priorities for a Planet in Transition* I compiled statements from various space visitors and their contactees pointing the way out of this predatory system that runs on human misery and natural destruction which we have somehow come to accept as inescapable facts of life. Some of these are worth repeating here as examples of manifesting our growth in consciousness in the way we relate to our fellow human beings. For instance, in 1964 George Adamski wrote: "[T]o have a healthy and prosperous society, that which causes the most trouble must be removed. As we all know, this stigma is poverty in the midst of plenty. It is the cause of sickness, crime, and the many evils that we know and when it is removed these bad results will vanish."[37]

After his contact experience in 1955 Truman Bethurum wrote he got "the impression that cooperation among all of their people is an inherent feature of their lives, and that poverty is unknown. Also that what we call riches or wealth is certainly more evenly distributed than on our earth."[38] George Adamski's experience on Venus confirmed this: "Their means of exchange is a commodity and service exchange system, without the use of money.

All production is for the benefit of everyone, with each receiving according to their needs. And since no money is involved, there are no 'rich'; there are no 'poor'. But all share equally, working for the common good."[39]

About the planet 'Iarga' (which I have elsewhere given reasons to believe is Mars[40]) that he was shown in the mid-1960s Dutch businessman Adrian Beers (writing under the pseudonym of Stefan Denaerde) said: "I understood that everyone here had equal rights. They lived in the same house, rode the same cars and stepped into the same trains. There was neither rich nor poor; there was no separation between nationalities, races or colors."[41] The prerequisites for such a system, his hosts explained, were universal freedom, justice and efficiency.

In this respect, it may be seen as a hopeful development that according to a BBC poll in April 2016 more and more men and women are identifying themselves as 'global' rather than national citizens, despite rising nationalism in richer nations. The poll, that was conducted among more than 20,000 people in 18 countries, came with the disclaimer that " 'global citizenship' is a difficult concept to define and the poll left it open to those taking part to interpret. For some, it might be about the projection of economic clout across the world. To others, it might mean an altruistic impulse to tackle the world's problems in a spirit of togetherness – whether that is climate change or inequality in the developing world." [42]

In many quarters of the modern world, the word 'spiritual' raises suspicion or it is seen as an excuse for inertia or as a hobby for self-indulging rich people. So for those who would otherwise be on board with the required changes in politics and economics, the word 'spiritual' may be off-putting. However, in his books through Alice A. Bailey the Tibetan Master D.K. explains that it is essentially synonymous with the expansion or growth of consciousness: "The word 'spiritual' does not refer to religious matters, so-called. All activity which drives the human being forward towards some form of development – physical, emotional, mental, intuitional, social – if it is in advance of his present state is essentially spiritual in nature."[43]

According to D.K. education is – or should be – a deeply spiritual enterprise which concerns the whole human being, including his divine spirit and, significantly, his definition appears on page 1 of the book *Education for the New Age* (1954). In it he also gives this unequivocal goal for education: "Two major ideas should be taught to the children of every country. They are: *the value of the individual and the fact of the one humanity*."[44] When a Master from Saturn addresses the group that hosts George Adamski on board a mothership, he emphasizes the same notion: "We will tell you of the physical life of other worlds, as well as what you call spiritual or religious truths, although we do not make that kind of division. There is but *one* life. That life is all-inclusive, and until men of Earth realize that they cannot serve or live two lives, but only one, they will be constantly opposing one another. That is one major truth that *must* be learned by all Earth men before life on your world can match life on other planets."[45]

In this light it is compelling to see how the notion of socioeconomic justice through sharing, or the establishment of right human relations, is brought up consistently in the wisdom teachings, beginning with Helena P. Blavatsky, who didn't hold back in this article from 1890: "Our age, we say, is inferior in wisdom to any other, because it professes, more visibly every day, contempt for truth and justice, without which there can be no wisdom. (...) Because this century of culture and worship of matter, while offering prizes and premiums for every 'best thing' under the sun, from the biggest baby and the largest orchid down to the strongest pugilist and the fattest pig, has no encouragement to offer to morality, no prize to give for any moral virtue... Because, finally, this is the age which, although proclaimed as one of physical and moral freedom, is in truth the age of the most ferocious moral and mental slavery, the like of which was never known before. ... Rapid civilization, adapted to the needs of the higher and middle classes, has doomed by contrast to only greater wretchedness the starving masses."[46]

Describing the emergence of international business after WWI in his very first public talk in 1922, when he was still being prepared as the 'mouthpiece' for the World Teacher for the new age, Jiddu Krishnamurti (1895-1986) said: "...international action is being forced upon us, though it is carried out in the wrong spirit, and for an immoral purpose, i.e., to exploit the poor, the needy, the suffering, and this method, if it continues, will inevitably lead to another war." As a solution he proposed the establishment of "an International Board, to control the commerce of the world, not for the profit of a chosen few, but for the entire world."[47]

Likewise, travelling around India on a quest for wisdom and truth in the early 1930s, British author Paul Brunton (pseudonym of Raphael Hurst, 1898-1981) eventually met Sri Shankara the 66th, with whom he had an interesting exchange about the political and economic conditions in the world. When asked when these conditions will begin to improve, Sri Shankara replied: "A change for the better is not easy to come by quickly... It is a process which must needs take some time. (...) Nothing but spiritual understanding between one nation and another, and between rich and poor, will produce goodwill and thus bring real peace and prosperity." When the author expresses his doubt if that is a realistic hope, his holiness replied: "The eyes of a patient man see deeper. God will use human instruments to adjust matters at the appointed hour. The turmoil among nations, the moral wickedness among people and the suffering of miserable millions will provoke, as a reaction, some great divinely inspired man to come to the rescue. (...) The process works like a law of physics. The greater the wretchedness caused by spiritual ignorance, materialism, the greater will be the man who will rise to help the world."[48]

American Manly P. Hall (1901-1990) became well known as a prolific writer and speaker on the subject of esoteric teachings and is perhaps best known for his encyclopaedic volume *The Secret Teachings of All Ages* (1928). In the Preface to its Diamond Jubilee Edition of 1988 the author reminisces the circumstances that led to its writing, when he had a brief career on Wall Street between the end of WWI and the Great Depression of 1929: "My fleeting contact with high finance resulted in serious doubts concerning business as it was being conducted at that time. It was apparent that materialism was in complete control of the economic structure, the final objective was for the individual to become part of a system providing economic security at the expense of the human soul, mind and body. (...) We are now coming to the end of the twentieth century, and the great materialistic progress which we have venerated for so long is on the verge of bankruptcy. (...) We were told that the twentieth century was the most progressive that the world has ever known, but unfortunately the progression was in the direction of self-destruction."[49]

In a treatise on esoteric healing the Master D.K. explained: "The keynote to good health, esoterically speaking, is sharing or distribution, just as it is the keynote to the general well-being of humanity. The economic ills of mankind closely correspond to disease in the individual. There is lack of a free flow of the necessities of life to the points of distribution; these points of distribution are idle: the direction of the distribution is faulty, and only through a sane and worldwide grasp of the New Age principle of sharing will human ills be cured..."[50]

Finally, while studying with some of the Masters in Tibet in the 1930s, Dr Murdo MacDonald-Bayne was told: "We can only live happily together when we are human beings; then we shall share the means of production in order to supply food, clothing and other necessities for all, without stint of selfishness."[51]

Based on his contacts with extraterrestrial visitors the Canadian contactee Wilbert Smith wrote: "Civilization is the relationship of beings with others of the same kind. This is something about which we know a little, but we don't do very much. We have a civilization of sorts but it isn't very good... As long as there is disagreement about how civilization should be set up, just that long will it continue to evolve. (…) [The] important thing is to recognize that what we have isn't very good and the ultimate is still far off, and we should work toward it."[52]

Nevertheless, George Adamski remarked: "All great teachers have taught the law of respect, love, and brotherhood. Jesus, whose teachings are the basis of every denomination of the Christian world, gave us one commandment… the commandment of love without judgment. Yet look at the divisions, resentments and hatreds prevalent among the people of Earth; all of which have laid the foundation for wars and rumors of wars confronting us on every side. If the people on other planets had lived their teachings no better than Earthlings, they, too, would be experiencing the same turmoil we find around us today."[53] In order to help us, a Master from Jupiter told him: "We are ready at all times to become servants unto the Earthly ones, guiding them if they but let us, to joys and happiness undreamt of by men on Earth. But the unification of thine household must come first. Thine right for servants must be earned. We too have much to give, but what we have, through lack of knowledge and wisdom, can be to the detriment of the children of Earth."[54]

That we, too, have the ability to grow and manifest our growth should not be doubted, according to a "profound Space Teacher" with whom Howard Menger had a meeting in august 1956: "Man is limited, but God is unlimited, infinite, expressing in all men, all forms. Men are student gods, going through a school of expression on this planet and many others, seeking knowledge and wisdom so that he may serve his brothers and the Infinite

Father of creation. Man continually progresses up the ladder toward perfection, and though a rung may break under the weight of his many errors, still his goal is to reach the top and one-ness with the Infinite Father."[55]

Humanity is One. That is, in the grander, planetary scheme of things the human race is an integral being that is approaching the point where its thoughts and actions become coordinated. The aim of life, therefore, says Benjamin Creme, is to represent "the unity which already exists because every atom in the manifested universe is interrelated with every other atom. Unity is not simply an idea which we can hold or not hold; it is driving us on our evolutionary process. This evolution, expansion of consciousness, must be a process of ever-widening awareness of unity and a synthesis of all the possible aspects of unity that exist until you have the 'Mind of God'..."[56]

This should not be seen as a 'hive-mind' where individuality ceases to exist, but as the coordination of individual propensities and talents, and the ability to cooperate on a sufficient scale to sway humanity's actions towards the progress and evolution of the whole. Until now, mankind has been at war with itself, all too often literally. With the coordinated cooperation of a critical mass of individual human beings – like brain cells in the mind of the planetary Being – humanity will firmly set foot on the path to fuller integration and integrity. Just as the individual depends on his mind for successful rational living, says Vera Stanley Alder, "The entire human family is also one being – but it is still scatterbrained. Our lunatic world must have a collective mind, sufficiently powerful to influence it, before present madness can end. The task that lies ahead for those who will embrace it is to GIVE THE WORLD A MIND!"[57]

As we broaden our view, we see that true spirituality is not a matter of simply meditating and closing our eyes to the suffering around us in order to feel more comfortable with ourselves, to use a popular oversimplification. Although it certainly requires introspection to eventually know oneself and thereby understand the world, true spiritual attainment – or growth in consciousness – is also given expression through what has been called 'love in action'.

However, this is not something that will descend on Earth with the appearance of flying saucers or, for that matter, the Second Coming. As George Adamski explained to a correspondent in 1950: "Babylon is beginning to shake on its foundations and the false gods can be heard everywhere, for the day of their doom is at hand. But for some time yet they shall appear as a power on earth and many will worship at their feet, for personal possessions will demand protection in order to survive."[58]

In a lecture in 1955 he points out: "…it states in our own Bible that in the latter days (as we might call it at the moment) that when these things will be happening … like 'signs in the sky and war and rumors of war' … we will have come to an end of a cycle, or as some people call it, a 'dispensation'."[59] About our own responsibility for bringing a new dispensation into being, he asked: "What chance would Jesus have if He were to return to Earth in fulfillment of Bible Prophecy? (…) Unless one's conscious perception is awakened, rather than sleeping under the blanket of materialism, how could one hope to recognize a man who in appearance would be no different from others? Were Jesus to return and be accepted, it would mean that all of our present systems would be overthrown to make way for His Cosmic Teachings. Are we prepared for this?"[60]

When UFO researchers read this, they would no doubt file it as evidence that contactees were mystics who made the sighting of flying saucers into a new religion – and many have indeed presented the contactees' accounts as such. After all, without any evidence of a space craft or a Divine Messenger and nothing to verify their claims, these were just stories for the gullible or the religiously desperate. However, as my research shows, the evolution of consciousness that the space visitors speak of is seen as a fundamental movement of life in both systems science and the wisdom teachings. In his 2006 book *The Chaos Point – The World at the Crossroads*, for instance, professor Laszlo also points out that the current time of crisis, in which our systems can be seen to fail and fall apart, is not the end of the world, but the end of a phase of the world after which a new world could dawn.

As professor Laszlo says elsewhere, "our consciousness evolves periodically in association with our body and incessantly in association with as well as beyond our body, and all this evolution is nonrandom and directional."[61] In this light we need to ask ourselves how 'religious' it is to think that over the aeons individuals like us have advanced to a state beyond the human consciousness – as postulated by the same Mme Blavatsky who was the first to assert the primacy of consciousness over matter?

That there have been such Teachers in the past is undeniable, when we see the enormous impact they have had on humanity, that continues millennia after they walked the Earth. Each of these Teachers, without exception, is sent into the world from what the wisdom teachings call the kingdom of souls, that kingdom in nature that is made up of those men and women who have completed the evolutionary cycle on Earth, but remain here to guide and inspire the rest of humanity, in service to the Plan of Evolution. Despite their humble initial followings, their teachings appealed to numbers so large that less evolved successors could only stay in control by adding rules and dogma that were never part of the original teaching. And to those who frown at the term 'Master' Benjamin Creme says: "They are Masters over Themselves. They are Masters in the sense that They have complete awareness and complete control on every plane of our planet. That is what makes a Master."[62]

It is informative here to read what Pierre Monnet learned from his space contacts: "It would be good if Earthlings stopped making the mistake of equating Jesus with religious dogma. Jesus did not come to create a new religion. Jesus came only to teach us Life. To teach us the universal cosmic laws that create, maintain and perpetuate Life. Jesus came to teach us the first cosmic universal law – the Law of Love. Love in simplicity. Jesus never wanted a religion to be formed around the teachings he brought us, and especially not around himself."[63]

As ritualized forms of spiritual practice, religions have seriously and demonstrably muddled up the origin of life, the early history of mankind, and even misinterpreted, misunderstood or simply mystified aspects of the life and mission of their respective Teachers. And although this has been the cause of much suffering, unnecessary conflict and even war throughout history, this in itself does not invalidate the original teachings.

During the course of my studies and research of the Ageless Wisdom teaching I have come to find some striking similarities in every major religion, which I summarized in chapter 1 as the cyclical revelation through (1) the coming or return of a Teacher at the beginning of every cosmic cycle, about (2) the source and evolution of consciousness, which (3) needs to be given expression by living according to the Golden Rule to establish right human relations.

The first correspondence, about the recurring appearance of a Teacher at the beginning of a new cosmic cycle, was explained by the Master D.K. as the esoteric Doctrine of the Coming One or the **Law of Cyclical Return**. In *The Secret Doctrine* Mme Blavatsky makes ample references to the Teachers of humanity throughout (known and unknown) history, each of whom brought a new revelation about the nature of reality – or God – and the way to reconnect with or achieve greater awareness of it. The names of many of these Teachers are well-known even now, such as Hermes, Hercules, Vyasa, Rama, Krishna, Gautama Buddha, Jesus Christ, and Mohammed. According to early Buddhist scholars Gautama already announced the Teacher for the coming age as follows: "At that period, brethren, there will arise in the world an Exalted One named Maitreya, fully Awakened, (...) unsurpassed as a guide to mortals willing to be led, a teacher for gods and men..." (*Digha Nikaya*, 26)

In their book *Presence – Human Purpose and the Field of the Future* systems scientist Peter Senge and three co-authors present a new theory about change to improve our world. Writing about their discovery of similarities to shifts in awareness, they found that while each religious tradition terms it differently – e.g. bliss, grace, opening of the heart, cessation of thought, "all recognize it as being central to personal cultivation or maturation".[64] As a student of the wisdom tradition, it is easy to see these "shifts in awareness" as the expansion of consciousness that is achieved when people live the teachings as outlined by the Teacher for that age, under the **Law of Evolution**.

When we live according to the precepts in the actual Teachings, by living the Laws of Life that come to us through cyclical revelation, we expand our consciousness and restore this awareness of our innate divinity as we integrate the various aspects of our being. But we can only take the next step when we manifest our expanded awareness by including the well-being of our fellow man, wherever he may be. Therefore, any true expression of this growth in consciousness must manifest itself in right human relations through the **Law of Harmlessness**, better known around the world as the Golden Rule (see Appendix II).

Perhaps the Laws enumerated here may be seen as some of the 'algorithms' that Ervin Laszlo says inform the evolution of coherent systems in the spacetime domain of our universe. In that case, we may ask ourselves how we might apply these Laws or use these algorithms to advance our evolution.

According to Dr Murdo MacDonald-Bayne, "the Universe follows an orderly plan of continuous progression fulfilling the Law or the 'will' of the Infinite which occupies the central place in the completed idea held in His all-embracing Mind... The most important part of this creation, to us, is ourselves and the use of our own creative power. There is only one Mind in which everything is and can ever be. We are creative in this one Mind, but, if we are unaware of this fact, we will not be familiar with the ways and means of *how* to create."[65]

In the words of Benjamin Creme, it is unity that drives us on our evolutionary process which, he says, "necessarily takes place, because it is part of the great outbreathing of the Creator... It is the Becoming of the Creator."[66] Elsewhere he adds: "God is an experience to be expressed. You expand your consciousness to become aware. It is all to do with awareness."[67]

As a metaphysical teacher with the Royal Order of Tibet in California in the 1930s, George Adamski taught: "How does man find himself? When he realizes the reality of life, when he realizes his true self, he then must step into a new garment and learn to say: 'Thy Will Be Done'." Here he refers to Jesus' statement "Not my will, but Thy Will be done", which means that as we grow in consciousness and increase our awareness of reality, we learn – or should learn – to align our individual goals and efforts with the plan of evolution insofar as we are able to perceive and understand it. "When we learn to work with the laws we let our free will be guided by the Thy-Will and we shall then begin to transform our whole being."[68]

Speaking through Dr MacDonald-Bayne the Master Jesus said: "When your consciousness becomes 'aware' of Life, Life – being the servant of all – shall manifest according to your awareness of It. Your consciousness unfolds through the realisation of the power of the Spirit; the consciousness then reveals and expresses that which the consciousness is aware of."[69] Eventually "you will not only see the work that is being done upon this planet, but you will also be in contact with forces, Spiritual Forces, which are working in other planets..."[70]

On this note, in answer to the question if there will ever be an interchange of peoples, ideas, and cultures between planets, Howard Menger's contact replied: "This is inevitable. You can delay God's plan, but never stop it. Interplanetary brotherhood for earth's people is dependent upon the degree of decline of hostility and the degree of increase towards tolerance, love, and good will toward our fellow men."[71]

So far, we have seen how the manifestation of life through the growth of consciousness as found in systems science, can be witnessed in the growth of social movements over the past nearly two centuries, as well as in the wisdom teachings and the accounts about the visitors from space. We have also seen how the lack of its manifestation increasingly threatens our physical, psychological, and spiritual well-being. The question now is if there is any evidence that the notion of our innate oneness is being expressed in the political, economic, and scientific fields, that we are in fact capable of giving expression to our growing awareness and include 'the other' in our experience of life, or that the 'algorithms' or 'Laws' that we identified earlier are indeed beginning to be applied?

'Treating the other well' was central to Martin Luther King's approach in his fight against injustice during the civil rights movement in the 1960s. In an interview with a Dutch newspaper during the massive Black Lives Matter protests that swept the US in the wake of yet another blatantly racial killing by the police, Dr King's press secretary Harcourt

Klinefelter recalls that MLK "truly believed that our weapon – radical love – was more powerful than the atom bomb. 'Because the bomb can only destroy', he would say, 'while love can change people's attitudes'."[72] A similar approach was taken by Nelson Mandela, who led South Africa from apartheid to a multiracial democracy without bloodshed, even if systemic injustice and inequality have not been rooted out yet.

The year 2019 saw a remarkable revival of Dr King's 'weapon'. In February, despite a ban on demonstrations, every Friday hundreds of thousands of protesters filled the streets of the capital of Algeria, until on 1 March an estimated three million people across the country stood up in protest against incumbent President Bouteflika after he announced his candidacy for a fifth term. And even when the president finally budged to protesters' demands and stepped down on 2 April 2019, they continued their peaceful protests. People had had enough and demanded that the structure of the corrupt regime would be dismantled after the entrenched elite had moved over, to allow a fresh start for the country after his twenty years in office. The movement was dubbed the 'revolution of smiles' because of its positive, peaceful energy and shared humanitarian goals, not least because it was supported "by hundreds of thousands of women across all age ranges and economic backgrounds to join the men on the streets, a place that is normally a male-dominated space".[73]

In June 2019 Gianmarco Negri won a mayoral election in the Italian town of Tromello against a far-right candidate of Italy's League party, that had come out on top in the European parliamentary elections the previous month on a platform of intolerance towards migrants and other minority groups. Mr Negri's victory was the more remarkable as he ran as an openly transgender candidate promoting kindness. Asked what message he had for League leader Matteo Salvini, he said: "That the politics of arrogance, violence and oppression, sooner or later will be overcome by a kindness revolution."[74]

When the opposition candidate in Istanbul, Turkey, had won the mayoral elections and Ekrem Imamoglu had been sworn in, Turkish President Recip Erdogan contested the election and cancelled the results to give his own candidate another go. Instead of turning angry or bitter, after three weeks in office Mr Imamoglu cheerfully accepted the President's dubious move: "We're game. With a smile on our face, with hope and love, and with respect we will take back what is rightfully ours." On 23 June 2019 Imamoglu won the elections again, this time with a majority of 800,000 votes. His strategy was inspired by Ates Ilyas Bassoy, whose gospel of 'radical love' helped the opposition break through the dominance of Erdogan's ruling AKP party in Istanbul and the capital Ankara for the first time in twenty years. Bassoy's philosophy says that it is impossible to fight evil with evil. While Turkey's leaders have been employing fear to rally large crowds, fear, according to Mr Bassoy, "is not the only powerful emotion. The other one is love, sharing, living together. And we say: don't be arrogant, do not alienate people, embrace your opponent's followers too."[75]

Popular uprisings are always a sign that the people no longer allow fear to inhibit their frustration at their lot, and beginning with the Occupy movement in response to the financial crisis of 2008, and the Arab Spring in 2010, the number of protests have been on a steady rise. A look around the globe in October 2019 saw countless marches, with people continuously taking to the streets in Iraq, Hong Kong, Lebanon, Spain, Chili, Haiti, Argentina, and Bolivia because "they are being robbed of their futures", according to professor of International Economics Glenn Rayp at the University of Ghent, Belgium.[76] And that is not even counting the massive climate strikes inspired by Greta Thunberg and her generation of secondary school students who have every reason to be gravely concerned about their future prospects with the current leaders' inaction on a despoiled planet. At the time of writing [August 2020], protests have toppled the government of Lebanon, and are besieging "Europe's last dictator" in Belarus, while in Thailand the young are leading the people's assault on absolute monarchy, demanding true democracy. If we see a majority of young people at such protests, perhaps we should, in the words of Benjamin Creme's Master: "Listen keenly to the young, they have the future safely in their hearts."[77]

While the reasons for the protests differ, the common denominator is a ruling class who are relatively indifferent towards the people and who fail in their primary role of organising society, says professor Rayp. An analysis on BBC News also found that "Many of those protesting are people who have long felt shut out of the wealth of their country. In several cases, a rise in prices for key services has proved the final straw."[78]

In an editorial comment about Brexit from October 2018 *The Observer* pointedly wrote that pro-Brexit politicians "do not understand how business is done these days, by multiple actors serving international clienteles, regardless of national borders. They do not see that on a planet of finite resources, sharing is a necessity, not a choice."[79] Ten years earlier, the financial crisis had convinced the late University of Manchester lecturer Mick Moran that "the officer class in business and in politics did not know what it was doing".

This is evident from the fact that global crises such as climate change and pandemics are driven, and in large part also caused, by the relentless pursuit of "economic growth", measured in Gross Domestic Product (GDP), that eats into our natural environment and disturbs nature's equilibrium. Countries, corporations and individuals with the highest purchasing or economic power, have the largest adverse impact on the crises. But the destructive impact from the crises affects those with the least economic power first and foremost – those who live in countries with the lowest incomes and the weakest protection against rising sea levels – who are often the ones labouring in sweatshops to produce the latest fashion items.

Moran formed a collective of academics who were dedicated to "exposing the complacency of finance-worship and to replacing it with an idea of running modern economies focused on maximising social good." Challenging the gospel of 'economic growth' because it denies a decent life to a large and growing number of people, they called themselves the Foundational Economy Collective to focus on bottom-up social regeneration for renewal and replenishment.[80]

Coincidentally – or not – the 'growth' gospel, as

expressed in the gross domestic product (GDP) was already challenged by the Gross National Happiness (GNH) index to measure the collective happiness and well-being of a population, introduced in the Kingdom of Bhutan in 2008. Another alternative to the GNP is the Happy Planet Index (HPI), introduced by the UK thinktank New Economics Foundation in 2006. If nothing else, these indices are indicative of a new way of thinking about the role of the economy as a function of society, rather than the idol demanding human sacrifice that it is now. That this is what our current socio-economic system is, will become abundantly clear now that Western economies are poised to go in freefall as a result of the corona crisis.

Fortunately, initiatives abound that point the way to a saner future for humanity. One such is the Wellbeing Economy Alliance (WEAll), a global collaboration of organisations, alliances, movements and individuals "working together to transform the economic system into one that delivers human and ecological wellbeing". A 'wellbeing economy' "starts with the idea that the economy should serve people and communities, first and foremost" instead of people and communities being at the mercy of large corporations hiding behind 'the economy' and 'the market'. The alliance advocates ten principles to "build back better" in the wake of the pandemic, among which socially just and ecologically safe goals, green infrastructure and provisioning, universal basic services and guaranteed livelihoods, fair distribution, better democracy, and cooperation.[81]

More explicitly, Share the World's Resources, set up in 2003, makes the case "for integrating the principle of sharing into world affairs as a pragmatic solution to a broad range of interconnected crises that governments are failing to sufficiently address – including hunger, poverty, climate change, environmental degradation and conflict over the world's natural resources." It calls for the implementation of the basic socioeconomic rights as granted in Article 25 of the Universal Declaration of Human Rights.[82]

Stop Ecocide was founded in 2017 to "support the establishment of ecocide as an international crime, in order to forbid and prevent further devastation to life on Earth."[83] In an open letter to the EU on 16 July 2020 Greta Thunberg and three other young climate activists align themselves with Stop Ecocide's aims, saying "We need to end the ongoing wrecking, exploitation and destruction of our life supporting systems and move towards a fully decarbonised economy that centres around the wellbeing of all people as well as the natural world."[84]

A framework for an economy that reflects our expanding consciousness to meet the needs of every man, woman and child within the limits of planetary capacity and environmental restoration is the 'doughnut' model created by economist Kate Raworth. Her Doughnut Economics quickly gained international recognition after its initial formulation was presented in a 2012 discussion paper by Oxfam. Soon it was being discussed in the UN General Assembly, the Global Green Growth Forum, as well as Occupy London. Mrs Raworth's 2017 book *Doughnut Economics: seven ways to think like a 21st century economist* has been translated into 18 languages. Needless to say that, in a world with nearly 8 billion people, this will require a massive readjustment among the better off segment of the world's population who have taken wasteful abundance for granted.

Of course, we did not arrive at this dire point of unsustainable social inequality and environmental destruction of a sudden. Scientists and visionary leaders have warned us for decades, such as in *The Limits to Growth* report published by the Club of Rome in 1972, and even in the 1960s there were scientists who expressed serious concern about the dangers of rising levels of CO_2 in the Earth's atmosphere. Through the political and economic systems that we elected and supported we, humanity, have chosen to ignore the warnings. As a result the situation has got to a point where we really have no choice but to change. In the previous chapter I quoted from several newspaper reports in which experts linked the Covid-19 crisis to our continued transgressions of nature's boundaries. For instance, the Global Footprint Network, who provide the Ecological Footprint metric that is based on United Nations data, says that our civilization is running at 40 per cent above Earth's capacity. Earth Overshoot Day is the date in any given year on which humanity's demand on ecological resources and services outpaces the Earth's capacity to regenerate over that period. In 2019, this was calculated to be 29 July. Tellingly, with Covid-19 shutting down society for several months in 2020, the world's demand overshot Earth's regenerative capacity a full month later, on 29 August.[85]

Already in 1948, the Master D.K. indicated that, "In the era which lies ahead, after the reappearance of the Christ, hundreds of thousands of men and women everywhere will pass through some one or other of the great expansions of consciousness, but the mass reflection will be that of the renunciation [of] the materialistic standards which today control in every layer of the human family. One of the lessons to be learnt by humanity at the present time (a time which is the ante-chamber to the new age) is how few material things are really necessary to life and happiness."[86]

The current situation is now so serious that, according to two theoretical physicists who specialize in complex systems in a paper in July 2020, "global deforestation due to human activities is on track to trigger the 'irreversible collapse' of human civilization within the next two to four decades." The report about their findings concludes with the pregnant notion: "So the most effective way to increase our chances of survival is to shift focus from extreme self-interest to a sense of stewardship for each other, other species, and the ecosystems in which we find ourselves. In other words, to avert collapse we either need to become ET, or spearhead a civilizational paradigm shift. Which is more probable? Ultimately, that's up to us."[87]

Having arrived here from the first two maxims that are inscribed on the column in the forecourt of the Temple of Apollo at Delphi – "Know thyself" and "Nothing in excess", can we see perhaps, the meaning behind the third maxim: "Surety ruins"? Isn't it our longing for security, our fear of change that inhibits us to act when faced with the results of our destructive habits? Isn't it fear which closes our minds to the new, and kills our spiritual hunger, while we are destroying everything our lives depend on?

The space visitors, too, have been warning us since the 1950s. For instance, George Adamski's contacts told him: "The old accepted thought patterns of people all over the world are changing rapidly. The under-privileged are crying for peace and equal rights with those who have enjoyed the good things of life."[88] And stressing the need to expand our consciousness to include our fellow humans: "If man is to live without catastrophe, he must look upon his fellow being as himself, the one a reflection of the other."[89] In the same vein, Brazilian contactee Dino Kraspedon learned: "Everyone is responsible for the misery and oppression in which [man] finds himself. If man changes his heart and makes up his mind to be merciful and good, he will at once have countless brothers at his side to help him, not to mention the help and joy from on High. Rest assured that the Father is more ready to give than the son is to ask."[90]

Albert Einstein's words, which I quoted before elsewhere, merit repeating in this context, which gives them a special relevance: "A human being is part of the whole called by us the universe, a part limited in time and space. He experiences himself, his thoughts and feelings as separated from the rest, a kind of optical illusion of consciousness. This illusion is a kind of prison for us, restricting us to our personal desires and to affection for a few persons nearest to us. Our task must be to free ourselves from this prison by widening our circle of compassion to embrace all living creatures and the whole of nature in its beauty."[91]

In *The Intelligence of the Cosmos* Ervin Laszlo expresses this notion in terms of systems science: "Natural systems evolve toward intrinsic as well as extrinsic coherence, and they acquire complexity in the process. The evolution of consciousness, in turn, is oriented toward the recognition of embracing oneness among the systems and the consciousness associated with the systems."[92]

This shows that the growth in consciousness that humanity has begun to demonstrate in declaring its highest aspirations, such as in the Universal Declaration of Human Rights, cannot be left to the realm of ideals with impunity. It has to be demonstrated in the way we organise society, the system in which we live, not only to progress further, but also because our survival now depends on it. As professor Laszlo puts it: "The sacredness of being human is that we each have a role in bringing the unfinished material reality into greater coherence, and thus completion. (...) In this case, we are each a conscious agent of cosmic realization and immanence. We each have an obligation in our existence on this planet to raise our individual, localized expressions of consciousness. In doing so, we both infect and inspire others in our lives to raise theirs."[93] After all, he says: "The hallmark of the evolved forms of consciousness is the mind-set that emerges in ethical, insightful, and spiritual human beings."[94]

As we grow in consciousness, we adapt and demonstrate our increased understanding, not only in the form of inventing new technologies, but also in the way we live and relate to ourselves, our fellow humans, and the planet. For if we don't advance on both ends in step, we will destroy ourselves – as we are rapidly learning from our experiences in the present circumstances on Earth.

This is the reason that contactees, without exception, are told about the need to demonstrate our awareness that the world is one, and humanity is one – that Life is one – in *actual practice*. And to educate humanity about this fundamental truth is the reason, as George Adamski was told, for the presence and the attempts of the Space Brothers to make themselves known without forcing themselves on us: "...our main purpose in coming to you at this time is to warn you of the grave danger which threatens men of Earth today. (...) Your people may accept the knowledge we hope to give them through you and through others, or they can turn deaf ears and destroy themselves. The choice is with the Earth's inhabitants. We cannot dictate."[95]

This was confirmed by many other contactees, including Pierre Monnet. When asked why his space contacts would not provide some physical evidence to help him convince people, he replied: "The sceptic is such that when you provide him with proof, he will demand ten more. When you bring ten, he will ask for a hundred. It never ends... In addition, it is essential to leave 50 percent doubt in the minds of the people on our planet. The commitment to evolution must come entirely from us. Without this, nothing that man undertakes will be of any value."[96]

However, the fact that the space visitors do not attempt to convince anyone against their will should not be used to ignore the available facts. In a criticism of the methods used by UFO debunkers, professor of Physics Auguste Meessen of the Catholic University Leuven, Belgium, said: "Either we give precedence to the facts observed and we try to explain them, even if that implies calling into question certain notions that seem to be well established. Or else, we give precedence to those notions, by immediately rejecting everything that upsets them."[97]

No matter from what angle we approach it, we need to take a broader view, and give expression to our expanding consciousness to establish right relations – with our true

Self, with each other, and with the planet. There are many instances that show a growing number of people realize we can only survive together, in cooperation. Thus far, these initiatives are still dispersed and therefore not yet effective in providing sufficient momentum to tip the balance towards justice and freedom for all and securing our common future on this planet.

As the state of the planet and humanity couldn't be more dire, we have reached the point where we have to make a final 'choice' – to step out of our limited individual awareness and expand our view of reality to align our individual free will with the Will (Purpose) of the Absolute, the Infinite, or God, depending on the nomenclature of your choice, with the algorithms of evolution – or to remain caught in separatism and proceed with ever fiercer competition and conflict, toward inevitable self-destruction.

As we have seen in this and the previous chapters, systems science now acknowledges what the Ageless Wisdom teaching has taught throughout human history: that consciousness is more fundamental than matter, and that it continuously evolves into higher expressions of itself. Significantly, we find the same notion of the evolution of consciousness at the heart of all religious teachings, as well as in the accounts of genuine contactees.

The foremost exponents and pioneers of this evolution on our planet are the Masters of Wisdom. As one to have worked most recently with one of these Masters, Benjamin Creme said that the Head of this group, the World Teacher, who is expected in every major world religion and whose personal name is Maitreya, has been ready to re-enter the life of humanity as guide and mentor since 1977.

He, Maitreya, the Eldest Brother of humanity, leaves no doubt that solving our current problems requires our direct involvement: "Many there are now who know this to be true, who desire to share, who long for brotherhood, yet act not. Nothing happens by itself. Man must act and implement his will."[98]

References

1 Robin McKie, ' "We are all Martians!": space explorers seek to solve the riddle of life on Mars'. *The Observer*, 12 July 2020. See <www.theguardian.com/science/2020/jul/12/we-are-all-martians-space-explorers-seek-to-solve-the-riddle-of-life-on-mars>

2 'Bad weather delays United Arab Emirates' ambitious Mars mission'. NOS Nieuws, 14 July 2020. See: <nos.nl/artikel/2340637-slecht-weer-vertraagt-ambitieuze-marsmissie-verenigde-arabische-emiraten.html>. Author's translation from Dutch.

3 George Adamski (1963), 'A Christmas Message to All Men of Good Will'. *Cosmic Bulletin*, December 1963

4 Jeff Bezos (2019), 'Going to Space to Benefit Earth'. See: <www.youtube.com/watch?v=GQ98hGUe6FM>

5 Franklin Foer, 'Jeff Bezos' Master Plan', *The Atlantic*, November 2019. See: <www.theatlantic.com/magazine/archive/2019/11/what-jeff-bezos-wants/598363/>

6 Adamski (1955), *Inside the Space Ships*, pp.238-39

7 Benjamin Creme's Master, 'The Great Mother', June 2001. In: Benjamin Creme (ed.; 2004), *A Master Speaks*, 3rd ed. pp.393-94

8 Adamski (1955), op cit, p.204

9 Adamski (1957-58), *Cosmic Science*, Series No.1, Part No.5, Question #100

10 Klee Irwin, Marcelo Amaral, and David Chester, 'The Self-Simulation Hypothesis Interpretation of Quantum Mechanics', 21 February 2020. See: <quantumgravityresearch.org/portfolio/the-self-simulation-hypothesis-interpretation-of-quantum-mechanics>

11 Annie Besant (1904), *A Study in Consciousness*, p.152

12 Vera Stanley Alder (1940), *The Fifth Dimension and The Future of Mankind*, pp.42-43

13 J. Krishnamurti, 3rd Public Talk, Madras, 5 February 1950. See <jiddu-krishnamurti.net/en/1950/1950-02-05-jiddu-krishnamurti-3rd-public-talk>

14 Adamski (1956), *World of Tomorrow*, p.5

15 Creme's Master, 'The Brotherhood of Man', October 2005. In: Creme (2017), *A Master Speaks*, Vol. Two, pp.47-48

16 Howard Menger (1959), *From Outer Space to You*, p.169

17 Adamski (1955), op cit, pp.93-94

18 Simon Tisdall, 'Covid-19 has changed everything. Now we need a revolution for a born-again world'. *The Guardian*, 24 May 2020. See: <www.theguardian.com/commentisfree/2020/may/24/covid-19-has-changed-everything-now-we-need-a-revolution-for-a-born-again-world>

19 Jonathan Watts, 'Bruno Latour: This is a catastrophe that has come from within'. *The Guardian*, 6 June 2020. See: <www.theguardian.com/world/2020/jun/06/bruno-latour-coronavirus-gaia-hypothesis-climate-crisis>

20 Adamski (1955), op cit, pp.94-95

21 Alice A. Bailey (1960), *The Rays and the Initiations*, p.746

22 Creme's Master (June 2001), op cit

23 Dino Kraspedon, *My Contact With Flying Saucers*, p.107

24 See: <www.brandt21forum.info/About_BrandtCommission.htm>

25 Charles Eisenstein, 'The Conspiracy Myth', May 2020. See <charleseisenstein.org/essays/the-conspiracy-myth/>

26 'Scourge of slavery still claims 40 million victims worldwide, 'must serve as a wakeup call'. UN News, 9 September 2019. See: <news.un.org/en/story/2019/09/1045972>

27 Institute for Policy Studies (2020), 'Global Inequality'. See: <inequality.org/facts/global-inequality/>

28 Daniel Fry (1954), *[A Report by Alan] To Men of Earth*. In: Fry (1966), *The White Sands Incident*, p.87

29 Kraspedon (1959), *My Contact With Flying Saucers*, p.89

30 May Bulman, 'Universal Credit linked to mental health problems for 63,674 people, study finds'. *The Independent*, 27 February 2020. See: <www.independent.co.uk/news/uk/home-news/universal-credit-depressed-mental-health-benefits-dwp-a9363776.html>

31 Patrick Butler, 'At least 69 suicides linked to DWP's handling of benefit claims'. *The Guardian*, 7 February 2020. See: <www.theguardian.com/society/2020/feb/07/dwp-benefit-related-suicide-numbers-not-true-figure-says-watchdog-nao>

32 Emily Stewart, 'The American unemployment system is broken by design'. *Vox*, 13 May 2020. See <www.vox.com/policy-and-politics/2020/5/13/21255894/unemployment-insurance-system-problems-florida-claims-pua-new-york>

33 Adamski (1949), *Pioneers of Space*, p.124

34 Mohammed Mesbahi (2020), *Towards a universal basic income for all humanity*, p.39

35 'Temporary Basic Income to protect the poorest people could slow the surge in Covid-19 cases, says UNDP', 23 July 2020. See: <www.undp.org/content/undp/en/home/news-centre/news/2020/Temporary_Basic_Income_to_protect_the_worlds_poorest_people_slow_COVID19.html>

36 Watts (2020), op cit

37 Adamski (1965), *Cosmic Bulletin*, December 1964, p.14

38 Truman Bethurum (1954), *Aboard a Flying Saucer*, p.137

39 Adamski (1957-58), op cit, Part No.1, Q18

40 See Gerard Aartsen (2015), *Priorities for a Planet in Transition*, pp.118-121

41 Stefan Denaerde (1977), *Operation Survival Earth*, p.38

42 Naomi Grimley, 'Identity 2016: 'Global citizenship' rising, poll suggests'. *BBC News*, 28 April 2016. See: <www.bbc.com/news/amp/world-36139904>

43 Bailey (1954), *Education for the New Age*, p.1

44 Ibid. p.47

45 Adamski (1955), op cit, p.208

46 H.P. Blavatsky (1890) 'The Dual Aspect of Wisdom'. *Lucifer*, Vol. VII, No.37, 15 September, pp.1-9. See: <www.theosociety.org/pasadena/hpb-sio/sio-dual.htm>

47 Besant (1922), *Theosophy and World Problems*, pp.75-76

48 Paul Brunton (1934), *A Search in Secret India*, p.127

49 Manly Palmer Hall (1988), *The Secret Teachings of All Ages*, Preface to the Diamond Jubilee edition

50 Bailey (1953), *Esoteric Healing*, pp.549-50

51 Murdo MacDonald-Bayne [n.d.; 1956], *The Yoga of the Christ*, p.137

52 Wilbert B. Smith (1969), *The Boys From Topside*, pp.39-41

53 Adamski (1957-58), op cit, Part No.5, Q99

54 Adamski (1949), op cit, pp.210-11

55 Menger (1959), op cit, p.175

56 Creme (2002), *The Art of Cooperation*, p.201

57 Alder (1940), op cit, p.224

58 Adamski (1962), *My Trip to the Twelve Counsellors Meeting That Took Place on Saturn*, Part 2, p.3

59 Adamski (1956), *World of Tomorrow*, p.1

60 Adamski (1957-58), op cit, Part No.1, Q94

61 Ervin Laszlo (2017), *The Intelligence of the Cosmos*, p.43

62 Creme (2012), *Unity in Diversity – The Way Forward for Humanity*, p.31

63 Interview with Pierre Monnet, *L'Ère Nouvelle*, January 2006. See: <pointdereference.free.fr/m/www.erenouvelle.com/PORTCO-7.HTM>. Author's translation from French.

64 Peter Senge et al (2004), *Presence. Human Purpose and the Field of the Future*, p.14

65 MacDonald-Bayne (1948), *What is Mine is Thine. How to Use Your Divine Power*, Part II, p.141

66 Creme (1997), *Maitreya's Mission*, Vol.Three, p.583

67 Creme, 'Questions and answers'. *Share International* magazine, Vol.35, No.6, July/August 2016, p.35

68 Adamski [n.d.; 1930s], 'Transformation of Body Consciousness', as reprinted in Aartsen (2019), *The Sea of Consciousness*, pp.27-34

69 MacDonald-Bayne [n.d.; 1953], *Divine Healing of Mind and Body. The Master Speaks Again*, p.90

70 Ibid, p.139

71 Menger (1959), op cit, p.169

72 Karel Smouter, 'Hoe de perschef van Martin Luther King naar Black Lives Matter kijkt'. *NRC*, 27 July 2020. See: <www.nrc.nl/nieuws/2020/07/27/hoe-de-perschef-van-martin-luther-king-naar-black-lives-matter-kijkt-a4007111>

73 Sara Benaissa, 'Hear me roar: The women behind Algeria's revolution of

smiles'. *New African*, 5 March 2020. See: <newafricanmagazine.com/22273/>

74 Angela Giuffrida, 'Italy's first transgender mayor says 'kindness revolution' can defeat far right'. *The Guardian*, 3 June 2019. See: <www.theguardian. com/world/2019/jun/03/italys-first-transgender-mayor-says-kindness-revolution-can-defeat-far-right>

75 Lucas Waagmeester, 'Turkse oppositie gaat Erdogan te lijf met "radicale liefde" '. *NOS Nieuws*, 11 May 2019. See: <nos.nl/artikel/2284169-turkse-oppositie-gaat-erdogan-te-lijf-met-radicale-liefde.html>. Author's translation from Dutch.

76 Lieven Desmet, 'Econoom Glenn Rayp: "Wereldwijd komen mensen de straat op omdat hun toekomst wordt gestolen" '. *De Morgen*, 28 oktober 2019. See: <www.demorgen.be/politiek/econoom-glenn-rayp-wereldwijd-komen-mensen-de-straat-op-omdat-hun-toekomst-wordt-gestolen~bee6aa3b/>. Author's translation from Dutch.

77 Creme's Master (2011), 'The ways of the New Time', April 2011. In: Creme (ed.; 2017), op cit, pp.159-60

78 'Do today's global protests have anything in common?', *BBC News*, 11 November 2019. See <www.bbc.com/news/world-50123743>

79 'The Observer view on the urgent need for a fresh vote on Europe'. *The Observer*, 20 October 2018. See <www.theguardian.com/commentisfree/2018/oct/20/second-referendum-hard-brexit-peoples-vote-way-forward>

80 Lynsey Hanley, 'What's the point of growth if it creates so much misery?'. *The Guardian*, 15 October 2018. See: <www.theguardian.com/commentisfree/2018/oct/15/we-can-rebuild-economy-foundations-up>

81 Wellbeing Economy Alliance, see:

82 Share the World's Resources, see: <www.sharing.org>

83 Stop Ecocide, see:

84 See

85 Earth Overshoot Day. See:

86 Bailey (1948), *The Reappearance of the Christ*, p.127

87 Nafeez Ahmed, 'Theoretical Physicists Say 90% Chance of Collapse Within Several Decades'. Vice.com, 28 July 2020. See: <www.vice.com/en_us/article/akzn5a/theoretical-physicists-say-90-chance-of-societal-collapse-within-several-decades>

88 Adamski (1961), *Cosmic Philosophy*, p.85

89 Adamski (1955), op cit, p.239

90 Kraspedon (1959), op cit, p.105

91 Albert Einstein, as quoted in *The New York Times*, 29 March 1972. See: <en.wikiquote.org/wiki/Albert_Einstein>

92 Laszlo (2017), op cit, p.46

93 Ibid., pp.61-62

94 Ibid., p.39

95 Adamski (1955), op cit, p.91

96 Interview with Pierre Monnet, op cit

97 Auguste Meessen, 'Le Phénomène OVNI et le Problème des Méthodologies', *Revue Française de Parapsychologie*, Vol.1, No.2, 1998, pp.79-102. Author's translation from French.

98 Creme (ed.; 1992), *Messages from Maitreya the Christ*, p.64

"Any sufficiently advanced technology is equivalent to magic."
—Sir Arthur C. Clarke, science writer

4. IN REALITY, CONSCIOUSNESS DRIVES TECHNOLOGY

In February 2015 NASA scientists were baffled by bright lights on the dwarf planet Ceres, in the asteroid belt between Mars and Jupiter. When asked, the Master of Wisdom whom Benjamin Creme worked with during his life confirmed these were Martian space stations.[1] Soon, however, in December that year, scientists proposed that Ceres may be home to an underground ocean and the lights in the Dawn probe's photographs might have been light reflecting off briny water-ice. On 10 August 2020 NASA's Jet Propulsion Laboratory (JPL) confirmed: "Mystery Solved: Bright Areas on Ceres Come From Salty Water Below" as scientists had figured out "that the bright areas were deposits made mostly of sodium carbonate…" The research "also found that the geologic activity driving these deposits could be ongoing"[2], and JPL's Julie Castillo-Rogez touted the discovery of brine-based minerals as a "smoking gun" for ongoing water activity, saying: "That material is unstable on Ceres' surface, and hence must have been emplaced very recently."[3]

If the lights on Ceres were the only anomaly to be found

in the solar system, the briny water-ice explanation would suffice to dismiss more prosaic explanations, such as "Martian space station" activity. However, as documented elsewhere, a growing number of anomalies are found around the solar system, including lights and cylindrical-shaped objects photographed by the Curiosity rover on Mars, and the frequent sightings of huge structures and radiant winged objects in photographs of solar activity taken by the Solar and Heliospheric Observatory[4], which NASA consistently brushes off as "interference" and other technical imperfections.

Yet, already in 1968 New Zealand airline captain and UFO researcher Bruce Cathie noted: "It is difficult to assess how much credence can be given to the many reports of strange objects seen near or passing across the planets during the past 200 years. Mars, Mercury, Venus, the Sun and the Moon, have all been associated with these events, and many of the world's astronomers have written of what they have seen." He says that we cannot dismiss all such accounts "because many have been seen by various observers and over the years similar sightings have been reported from all parts of the world."[5]

In July 2020 mysterious plumes were seen rising up from the surface of Mars by the Visual Monitoring Camera on board the European Space Agency (ESA)'s Mars Express. Clouds have been spotted on Mars as early as 1995, and the first "protrusion" in the location of the current plumes was recorded in November 2003.[6] According to ESA, "the cloud is made up of water ice, but despite appearances, it is not a plume linked to volcanic activity. Instead, the curious stream forms as airflow is influenced by the volcano's 'leeward' slope – the side that does not face the wind."[7] Equally unexpected was the discovery of phosphines in the atmosphere of Venus early September 2020, about which MIT professor Sara Seager writes: "This spectacular discovery is simply astounding. (...) We can argue there is no plausible way for PH_3 to be produced at the detected levels via any known atmosphere, surface, or subsurface chemistry. We are left with the incredible possibility that PH_3 might be produced by life."[8]

The mysterious activity of matter at the quantum level and the non-locality of consciousness require that we take into account the 'deep dimension' which systems science says informs objective reality, or the etheric planes of matter where, according to the wisdom teachings, the energetic blueprints of physical reality may be found. Likewise, these space mysteries would demand the same broadened approach as we have applied to the UFO phenomenon and the question of life outside Earth. And if mainstream science itself concludes that life is more likely the rule than the exception in the universe (pages 13-14), there is no reason why such space anomalies could not be the deliberate physical plane effect of the activity – invisible to our eyes and probes – of space brothers and sisters, perhaps with the aim of expanding their range of attention-grabbers to let us know that there are more things in heaven and earth than are acknowledged in our astrophysics or government statements.

That there *are* more things in heaven and earth has never *not* been known but has been increasingly ignored and denied, ironically as a result of the 'Enlightenment', by fundamentalist materialism (also known as 'scientism'), especially in the 18th and 19th centuries, to the point that telepathy, clairvoyance, and other extrasensory abilities are still largely frowned upon by science. Yet, Vera Stanley Alder says, "Man contains within himself an expression of everything, great and small, within the universe. He is subconsciously aware of this, and has, throughout his history, externalized into objectivity or physical manifestation, one by one, all those things which he embraces within himself." For example, she says, mankind has externalized "every muscle and sinew in his body in the form of tools, machines and engines. He has externalized his eyesight in the form of camera and cinema, his hearing in the form of music, telephone and wireless... His achievements are incessant, untiring and astonishing." She rightfully points out that "All the great minds [such as Pythagoras, the Buddha and the Christ] have declared man to be possessed of 'superhuman' powers which it is his duty to develop, and by means of which he learns to know and to control the hidden powers and entities and individualities of the invisible dimensions."

But, she continues, "Man appears to have shelved his inner life while he gives attention to the conquest of matter. He has left the understanding of himself and his own faculties to isolated communities such as the spiritualists... They and others have established the genuineness of clairvoyance, of prophecy, of healing, of phenomena in the realm of miracles and of extraordinary knowledge possessed by the sub- or super-conscious mind."[9]

Indeed, when an interviewer asked French contactee Pierre Monnet if compiling all our knowledge and scientific power would help us reach perfection, wisdom, and love, he replied: "Like many people, you put the cart before the horse.

You should know that the scientific knowledge of men only appeared because of the loss of the colossal psychic powers which they initially held. Man lost the powers with which he could do everything effortlessly, his mind creating matter. This loss is due to the negativity created by the misuse of free will. From that moment and very quickly, it needed to be compensated to be able to survive."[10]

We may wonder if this is the reason why the space visitor who contacted Enrique Barrios told him: "If you were aware of yourself, as you are of your surroundings [the material world], you would discover many things..."[11] Or in the words of Daniel Fry's contact: "Mankind (...) no matter where or when he may come into being, is endowed with the innate realization that there is an infinite intelligence and a supreme power which is greater than man's ability to comprehend. During the many stages of his development, man's attitude toward this power may vary from fear and resentment, to reverence and love. But he has always had the instinctive desire to learn more of the spiritual side of his nature and the creative sphere of this power."[12]

The occult law says that "everything unseen must become seen", and so humanity at large continues the work of externalizing its own inborn abilities, according to Ms Alder in 1940: "With amazing cleverness they have produced television, by means of which it is possible to see things happening at a great distance" while there have been "numerous attested cases of people who, while under hypnotic direction, were able to project their sight to a distant town and correctly describe what was taking place there."[13] One such case is physicist Russell Targ, who worked at Stanford Research Institute (SRI) investigating psychic abilities for the CIA, Defense and other US intelligence agencies, as well as NASA, during a 23 year programme. In a talk about psychic abilities in 2013 he cites many impressive experiments and cases that he worked on. He calls remote viewing a natural ability that can be trained to describe and experience what is happening at a distant place, or in the future, and references a "vast Buddhist lore" about psychic abilities. Remote viewing, he says, "pertains to the fact that we live in a non-local space-time, described scientifically most recently by Schrödinger in the 1920s and then proved in the 1970s and the 1980s. So the idea of non-local connections is *not* 'new age'."[14]

Also, in the Galileo Commission Report Dr Walach cites a study from 2018 which concluded that about 1 percent of the population possess the capability or sensitivity to perform remote viewing.[15] If nothing else, this shows how these 'paranormal' abilities are gaining acceptance, or are at least being seriously researched, among scientists and medical experts with an expanded view of reality.

In the second half of the 1990s the promise of the burgeoning internet beguiled many to expect the rise of a 'new consciousness' as the world wide web enabled everyone on the planet to be connected with everyone else.

Twenty-five years later and 'Big Tech' rules the web, and the world, for financial gain. They often depend on atrocious labour circumstances, where unhealthy and degrading working conditions, unforgiving production targets and the lowest possible wages push workers to the limits of economic destitution, and frequently, suicide. Anyone who reads the newspapers with some regularity will have seen headlines reporting such circumstances in Chinese factories for Apple computer and gadget components[16], Amazon.com warehouses in the USA, UK, and Germany[17], and cobalt and coltan mines in Africa where the minerals used in smart phones and tablets are delved.[18] All the while, despite or because of their massive profits and customer bases, big tech companies are unable or unwilling to stop the misuse of people's personal data collected through their platforms for undermining the democratic process through mass manipulation and fake news.

So, even though the internet and its technology can be seen as the outer manifestation of our interconnectedness on the inner planes, in its present form it comes at huge costs to the environment, workers and society. That said, it is not that the internet is evil, writes Alan Rusbridger, it "simply amplifies who we are", or rather, how a limited view of reality directs our behaviour.[19]

Living in a system where competition for material success and physical objectification are built into our structures and conditions our attitudes, we allow technology to amplify these symptoms of the ultimate externalization of our inner lives. As Enrique Barrios' contact told him: "The problem is not in the people, but in the systems they use. People have evolved, but systems have remained backward. Bad systems make good people suffer. These systems make people unhappy, which finally turns them bad. A good system of global organization can easily turn bad into good."[20]

On 15 August 1972 US radio journalist Bryce Bond had joined UK flying saucer enthusiast Arthur Shuttlewood for a UFO hunting mission on Star Hill, near Warminster,

England after sunset. The pair first spot a "triangular-shaped flying object, surrounded by coloured patterns, crossing the night sky with peculiar manoeuvres", disappearing as suddenly as it came. Shortly after, they noticed a glaring white object that danced in the air: "Suddenly, I heard a noise. It seemed as if something pushed down the wheat. (...) In front of my eyes, I could see a great imprint taking shape. The wheat was forced down in a clockwise direction. 'It' somehow had the shape of a triangle with a diameter of about 23ft (7m)."[21] What they had just witnessed was nothing less than the birth of a modern mystery – the crop formation phenomenon.

Although the more spectacular designs among the formations that are found every year in fields all over the world, particularly in south west England, will still make the columns of newspapers from time to time, crop circles seem to have become 'one of those things' for the media and the general public. Yet, one of the earliest books written about the phenomenon, *Crop Circles – Harbingers of World Change* (1991), shows that some of the early crop circle researchers were astutely aware of the dramatic implications of their mysterious appearance.

In the Introduction, editor and contributor Alick Bartholomew writes that "public confidence in our political and economic institutions is rapidly failing, as the hollowness and hypocrisy of 'the system' is exposed. We don't need to look for scapegoats; our leaders don't seem to be aware of what is happening, but we get the leaders we deserve, so it is we who have to change. It is just that we have lost our way, as 'the elder brother' is trying to say."[22]

Even in the early days of crop circle research and

studies, many noticed that they appeared on what some call 'ley lines', which are said to demarcate earth energies. In her book *Crop Circles. The Greatest Mystery of Modern Times* UK author and photographer Lucy Pringle makes various references to "energy lines" and writes: "It would appear that the large majority of both genuine and hoaxed formations can be found on natural energy lines. That energy lines affect us more than we realize, due to their electro-magnetic emissions, is still not generally recognized."[23] Perhaps it should be noted here that researchers distinguish between genuine and hoaxed crop formations based on several crucial differences. For instance, in genuine formations the stems of the crops, even fragile oil seed rape stalks, are always intact, in contrast with the stems being broken in formations that are made with ropes and boards. Another characteristic of a genuine formation is the occurrence of 'growth nodes' in the stems, as if the crop had been exposed to a burst of microwaves.[24]

With regard to 'energy lines', in the 1960s captain Bruce Cathie discovered a regular pattern along which UFO sightings were taking place around the world. Beginning in the summer of 1952, captain Cathie was witness to a number of UFO sightings and as an airline pilot he began to notice some regularities in the flight paths where they showed up. In 1965 he came in contact with the New Zealand UFO organisation run by Henk and Brenda Hinfelaar, who in 1959 had organised George Adamski's tour of the country as the first leg of his world lecture tour. They provided Cathie with access to a meticulously indexed collection of UFO sightings from 25 different countries over a period of twelve years. Along with his own observations he found that

New Zealand, "and probably the whole world, were being systematically covered by some type of grid system".[25]

Cathie came to believe that the significance of his discovery lies in the grid-like pattern which he plotted on the basis of sightings by himself and many others. His research also convinced him that it was not the space visitors we need to fear, but the powerful forces behind the UFO cover-up and the systematic disinforming of the public about it.

Based on his research, captain Cathie concluded: "Even while you read this, interplanetary space-ships are rebuilding a world grid system... When I say re-building, I mean exactly that, because my investigations show that an earlier grid existed way back in history."[26] He assumed the space visitors use this grid to draw power from for propulsion, and perhaps for navigational purposes. However, according to Benjamin Creme, who worked with the Space Brothers in the 1950s, the pattern that Bruce Cathie uncovered in his research of UFO sightings has a somewhat different purpose: "What the Space People are doing in the crop circles in particular is recreating to a certain degree the 'grid' of our earth's magnetic field on the physical plane."[27]

According to Creme the crop circles are made by energy: "The people in the spaceship form a design in their mind, simple or complex, and then, with a combination of technology and thought, that design is created in the crop." This, he says, is done in a matter of seconds. "They always use a universal system based on '9', rather than '10', which we will adopt in the future. When we understand the nature of the mathematical ratios in the true sense, we will begin to understand the crop circles better."[28]

Assuming the grid served as the power source for

UFOs, captain Cathie used it to predict not only where UFO sightings might take place, but also where the testing of nuclear bombs could be expected. This wasn't too far off the mark, given Benjamin Creme's explanation that UFOs are often seen in places with high concentrations of nuclear radiation which they neutralise. According to his Master, "Earth scientists are confident that they have, indeed, tamed the monster, and can keep it under control. They do not realize that their instruments (…) measure only the lower aspects of nuclear radiation, that stretching above these dense-physical levels are levels finer and more dangerous to the health and well-being of all. But for the tireless efforts of our Extra-planetary Brothers in assuaging this invisible peril in so far as the karmic law allows, our plight would be perilous indeed."[29]

Captain Cathie, too, had a good inkling of the dangers inherent in nuclear fission technology: "I am convinced that a completely safe atomic-type reaction could be maintained over any point on the earth's surface if the harmonic frequencies of the UFO grid were utilized. No atomic disruption of material particles would take place, and the whole business would be completely under control. It is not necessary to dismantle atoms in order to use the forces binding them together. A more subtle process of conversion is indicated by our visitors."[30]

Confirmation of this conviction can be found in Adamski's *Pioneers of Space*, when the party from Earth was shown a central heating unit which serviced a settlement on the dark side of the Moon. Suspecting it ran on atomic power, Dr Johnston inquired if "the power used for heating here is similar to what we have on Earth, but which so far we have never used in this manner." The Moon scientist answered: "...what you Earth men have now is very dangerous to work with, for you really haven't got anything except wild, uncontrolled power. Your job is taming this power down so that it will not be injurious to anyone. You will have to get another element to do this with, as we have done. Then you will be able to use it so well that even a child can handle it without any harm. You have some of that element on Earth, but you have not yet found it. It is through the proper use of this power that the dwellings on the dark moon are being heated."[31]

This assertion is confirmed by Benjamin Creme's Master, who says that led by greed humanity has neglected "a perfectly safe alternative use of the energy of the atom. Atomic fusion, cold and harmless, could be theirs from a simple isotope of water, everywhere available in the oceans, seas and rivers, and in every shower of rain."[32] Even in 1923 the Master D.K. wrote: "When the energy of water and of the atom is harnessed for the use of man, our present types of factories, our methods of navigation and of transportation, such as steamers and railway apparatus, will be entirely revolutionised."[33]

Interestingly, in a research paper published in *Nature* in July 2018, a group of scientists from the California Institute of Technology, the Free University of Berlin and the ESA describe a method involving "photo catalysts" to split or recombine water molecules that may be used to produce fuel, air and water during deep space flights. Normal electrolysis is already used to make hydrogen for fuel cells, but requires heavy equipment that is not suitable to build into a space module. If scientists can make this new technology work they could effectively use water as rocket fuel and transport 'fuel' into space in the form of water which, according to one commentator, "would be much safer than putting explosives on top of a fire tower powered by more explosives."[34] Coincidentally, as George Adamski noticed on a Martian ship, "They even have an emergency water-maker. It seems that this water-maker catches the air or atmosphere, similar to a ventilator and then condenses the air into fresh water."[35]

About the use of our current nuclear technology Benjamin Creme later remarked: "The advice of Maitreya and the Masters will be to close down immediately all nuclear-fission power stations in the world. They could be replaced tomorrow with a safe, fusion process of nuclear power as an interim measure"[36] before a new type of power is available.

As we have seen, our current technology is really the extension, on the physical plane, of our innate abilities, or a compensation for losing these. In his *Treatise on Cosmic Fire*, which he says provides "the psychological key to the Cosmic Creation", the Master D.K. discusses 'fire' on three planes – the physical, the mental and the spiritual – as a system of philosophy which will link both spirit and matter. The origin of fire on these respective planes are the electrical fire-by-friction of the physical sun; the solar fire from the Heart of the Sun; and the electrical fire from the Central Spiritual Sun. In each of these, according to Benjamin Creme, lies a great mystery. "At the core of that mystery is the nature of light and therefore the nature of life itself." Under the stimulus of the Masters, he says, "humanity is approaching the point where we will begin to investigate the nature of life which demonstrates as the light of the sun. We have been told that Maitreya [the World Teacher], will eventually introduce a new technology, the Technology of Light. That will allow us to understand and use the secrets hidden in the nature of electricity."[37]

Here, Creme refers to the new technology that, he says, is directly linked to the crop circle phenomenon. The vortices in the electromagnetic field on which the crop circles appear, "will become 'batteries of energy' for humanity in connection with a new energy source, part of the science of light which lies just ahead for humanity."[38] He explains: "The Space Brothers are creating these vortices of power on the physical plane. A crop circle is really an outer sign of a vortex. A replica of our planet's magnetic field is being placed all over the physical world, not as huge and powerful as the planet's magnetic field, but powerful and widespread enough to be the basis of the Technology of Light. The light will come directly from the sun and be fused with the magnetism of the magnetic field. That will give humanity every form of power it needs."[39]

Elaborating, he says: "It is a gift for the coming time, the New Age which we are entering, which we call the Age of Aquarius, a time in which tremendous new discoveries in the use of the energy of the planet and of the sun will be found and adapted. This will entirely change our way of life on this planet, and will lead to a control of universal energy such as we cannot begin to imagine."[40]

In light of these statements it is remarkable to see some of George Adamski's observations about the use of light in *Pioneers of Space*: "On the instrument panel of the ship is a small screen upon which the light of the planet to which they are going is focused. The details of this instrument we don't know, but the principle is the same as that used by an astronomer when he gets an image in the dead center of the mirror of his telescope. Once the planet is perfectly focused upon the screen, one of the pilots pushes a button and the ship is definitely on the light lane towards its desired destination." And, "once this is accomplished, no outside interference can cause the ship to change its course for it is actually being pulled forward by the light."[41]

The end of the light beam would indicate the destination from which the light was focused or reflecting. This "highway of light" is the "drawing power by which this ship is travelling through space" at terrific speeds.[42] His hosts later explain that "the light does not go away from the planet into space, but rather it comes from out of space towards the planet. It is travelling towards the planet instead of from it. (…) That which you call light is radiation from the sun and is actually travelling from the sun to the planet."[43] Benjamin Creme's Master seems to refer to this when he says that "the light of the sun will become the fuel".[44] Most likely this does not involve the physical plane light of the electrical "fire-by-friction" that falls within our vision, but light of one of the two 'higher' fires mentioned by D.K.

Howard Menger's contact from space told him that the force of the space ship in which he arrived "will be difficult and probably impossible for you to understand. It is an electromagnetic force, not unlike the force which holds planets, suns, and even entire galaxies in their orbits. This force is a natural law, which has been given to us by our Infinite Creator to be used for good purposes." Asked why our scientists had not yet discovered this power, the answer came: "Ah, but they have. Yet they do not know how to apply it. If they did know the secret they probably would use it for destructive purposes. Until they are ready to utilize it for peaceful ends, our Infinite Creator will prevent their understanding it."[45]

In his book *The White Sands Incident* US contactee Daniel Fry, who had his first contact experience in 1950 but did not make it public until 1954, explains that the visitor he spoke with emphasized the need for humanity to discover "the utter simplicity of the basic laws or facts of nature. Then you will easily be able to produce effects which now seem to you to be impossible." He then gave an example: "When your engineers design a vehicle for transportation of freight or passengers, they feel it necessary to provide a means of producing an energy differential within the vehicle itself as a motive power. Yet, your ancestors, for thousands of years travelled to all parts of your planet in ships which had no internal energy source but which were operated entirely by the kinetic energy of the atmosphere..."[46]

Adamski and his fellow travellers were also shown an experimental power plant on the Moon with gadgets resembling parasols and others shaped like horseshoes on the roof. One of the scientists who escorted them through the plant explained that "these gadgets act as condensers, drawing out of space what we call static electricity. They call it energy [used only for alloying metals and research work]. (...) In fact they were experimenting at this very time with the idea of drawing more power out of space."[47] While the gadgets that Adamski saw on the roof of the plant were large, it is interesting that scientists at the University of Massachusetts Amherst recently described a method to harvest energy from the atmosphere too, but through minuscule protein wires. In a research paper published by *Nature* in February 2020, the researchers describe how they have developed a technology to generate energy from the environment that offers "a continuous energy-harvesting strategy that is less restricted by location or environmental conditions than other sustainable approaches".[48] Here again, if his book were fiction Adamski's foresight is striking, but the evidence gathered in these chapters supporting the likelihood of its authenticity is undeniable.

The Master D.K. said that one aspect of electrical phenomena produces cohesion, while another produces light. Although D.K. does not call it that, Benjamin Creme believes that the Master is referring to the Technology of Light when he says: "One of the imminent discoveries will be the integrating power of electricity as it produces the cohesion within all forms and sustains all form-life during the cycle of manifested existence. It produces also the coming together of atoms and of the organisms within forms, so constructing that which is needed to express the life principle."[49]

It is noteworthy in this context that the power source on Mars is described by Adamski's hosts as "electrical where it is repulsive as well as attractive, composed totally of a positive force... The only thing we know is that it is composed of seven different major elements that govern the solar system, and each tiny atom is made up of these seven elements."[50]

This is reflected when, speaking of Fohat as "the *cause*

and the *effects* of Cosmic Electricity", H.P. Blavatsky writes that these seven elements, in occult terminology, are "*the seven primary* forces of Electricity, whose purely phenomenal, and hence *grossest* effects are alone cognizable by physicists on the cosmic and especially on the terrestrial plane."[51] These effects seem to be what Benjamin Creme referred to when he said: "All matter is a precipitation of light and so the field of matter is light precipitated into seven more or less material planes."[52]

Elsewhere Blavatsky stated that Fohat, "running along the seven principles of Akasha, acts upon manifested Substance, or the One Element, (...) and sets in motion the law of Cosmic Evolution, which, in obedience to the Ideation of the Universal Mind, brings into existence all the various states of being in the manifested Solar System."[53]

In this respect Benjamin Creme says, "When we switch on the electric light we are actually tuning into the lowest, physical-plane, level of the light of the world. (...) When we understand the true inner nature of electricity, we will gain control of the forces of the universe. We have only, as yet, touched the surface of electricity. We handle it more and more adeptly as time goes on; we can make it do all sorts of things. It has created our technology. We use it for heat, light, the creation of motion, and so on. But still we have missed the inner mystery of electricity. That, when discovered, will lead us to the control of light itself."[54]

Beginning in 1997 a new phenomenon started being observed. This involved the appearance of 'circles of light' reflected on walls or pavements from the sun shining on otherwise rectangular windowpanes. Where before the reflection would match the shape of the pane, now their reflections had changed into circles, usually with an 'X' inside with a large, very bright centre. While it seems that some types of glass naturally reflect light in a similar pattern, the phenomenon observed here involves a regular reflection that has changed, without the windowpane having been changed. This phenomenon was first reported in Burlington, Vermont, USA, but quickly spread and can now be found around the world in many different shapes. According to Benjamin Creme's Master these light images are created by the World Teacher in cooperation with the Space Brothers.[55] Although these circles have a completely different function than the crop circles, Mr Creme added

Reflections from four rows of windows on the building and street opposite the office building behind the photographer, in Amsterdam, the Netherlands show circles and other patterns of light in the rightmost column, next to regular reflections in the six columns to the left. (Photo taken by the author, 29 October 2006.)

later, the phenomenon is "part of the preparation for the Technology of Light. It is an illustration of the Space Brothers' command of light phenomena" and that "if we understood the technology that created the circles of light phenomena we could rid the world of global warming in a very short space of time".[56] The circles are "energetic in nature and each pattern grounds an energy"[57]. On another occasion he said that they also have a healing dimension.[58]

Another phenomenon has been known about for much longer. 'Energy vortices' are anomalies that can be found across the North American continent. They are often open to visitors as roadside attractions with names such as 'The House of Mystery", 'The Mystery Spot', 'The Oregon Vortex', et cetera. Inside, marbles will roll uphill, objects will balance themselves with ease, and people report feeling lightheaded or top heavy. Two people standing opposite each other may be the same height, but when they switch positions one looks considerably taller than the other. I first read about this phenomenon – about the vortex in Sardine Creek, Oregon – in *Fate* magazine of July 1951, which also carried George Adamski's second article about photographing flying saucers.[59] In addition to the one in Oregon, in his book *Harmonic 695* Bruce Cathie also mentions one near Santa Cruz, California, and includes photographs as well.[60] Although sceptical, the Roadside America website lists no less than 37 of such vortices around the United States and Canada.[61]

When a reader described his experiences in a vortex at Columbia Falls, Montana in a letter to the editor of *Share International* magazine in 2001, Benjamin Creme stated that "the energy vortex was created by Maitreya. The force used is not magnetic but another energy in connection with a future technology. The energy field does affect our experiences of space and dimensions."[62] It remains unknown if the "future technology" that Mr Creme referred to in his comment is the Technology of Light. It is likely, though, since Creme and his Master have said that the Technology of Light that awaits us will be used for all our needs, domestic, industrial as well as for transportation.

Setting the demonstration of humanity's innate oneness as a prerequisite, in October 1999 the Master gave an evocative preview of its implementation: "Imagine cities of light lit by Light Itself; nowhere to be found the squalor and deprivation of today; imagine transport, fast and silent, powered by light alone, the far-off worlds and even the stars brought within our reach. Such a future awaits the men and women who have the courage to share."[63]

Indeed, with regard to transportation on Venus George Adamski observed: "For overland transportation they have a system similar to our railways, but since they use light for energy, they need no rails for their cars as we do. The power engine pulls thirty to forty cars, what we would call cars, yet they don't look like any that we have. They seem to be gliding right over the surface of the ground. In other words, they have all the modern conveniences so far ahead of us that we haven't even begun."[64] Later they find out that all power "is developed by the planetary government and is freely distributed to all places for all needs. It is generated from light, the same as that used by the airships." They also have a standby so that "should something happen to the apparatus generating their power from the light, they could resort to this sort of power, which was utilizing static electricity out of space."[65] Also on Venus, Adamski noted, "We haven't seen any large manufacturing plants of any kind or anything that would suggest it, but they tell us there are many, but the manufacturing places, we would call them factories, are built just as beautiful as homes or business buildings, so we couldn't tell them apart."[66]

On the planet that Dutch contactee Stefan Denaerde was invited to learn about, he observed: "The fully automated robot rail transport system operated with frictionless efficiency, moving vehicles of many sizes and configurations at high speeds. There were individual cars for small groups, collective units like trains for mass movement, cargo units for commerce, and even a peculiar development something like a hotel tram. A group of people wanting to travel together would order a unit that was fitted out as a self-service hotel and simply go where the mood took them. The system was marvelously efficient and could move over one million persons per hour past any point using only the upper six-lane rail system between the house blocks. The rail system and equipment was designed with a useful life expectancy of 1,000 years, a kind of [product] quality undreamed of on Earth."[67]

When the Technology of Light becomes available, we will also see major changes in the way artifacts for daily use are manufactured. According to Benjamin Creme, "More and more, we will develop machines which will create the artifacts which we use in everyday life, so the need for human beings in factories will increasingly disappear. Then will come a time when these same technological inventions will be created by mind, programmed by mind, with a very clear end-result already foreseen by the creator of the technology." The Masters themselves also create all kinds of instruments and apparatuses that do Their work and which they use in experiments: "The Masters do not think that They know everything. They are experimenting all the time, exploring the planet, experimenting with energy, with what happens when energy, for example, leaves in the form of a thought made by one person and is picked up by another. All of this is under constant experimental attention by the Masters. We ourselves will develop, more and more, this kind of approach."[68]

While manufacturing through the power of thought will initially cause concerns about employment, Benjamin Creme's Master says: "Without doubt, the main reason for increasing unemployment is the discovery and application of the new technology. More and more, the robot is replacing man in the more complex manufacturing processes. No man can compare, in speed of operation and repetitive accuracy, with the sophisticated machines now in use. This is as it should be. Many may lament the loss of human skills earned through long apprenticeship and training, but man is born for higher and more worthy efforts. Why should men compete with mere machines?"[69]

Here, we may again quote George Adamski, who said, "if we continue to move as we have been in the field of technology we will have machines or robots doing the labor of man as they do on other planets today. And this creation by man will have to supply his needs and give him time to develop a better world and himself."[70]

On one of the motherships on which he travelled, there "was a robot instrument which I was cautioned not to describe. I had noticed a miniature version of this robot in the Scout."[71] "Each pilot room has a robot. These, working singly or together, can fully govern the course of the ship, as well as warn us of any approaching danger."[72] On their tour of a Martian ship while on the Moon, Adamski describes one instrument that was explained to them as "an all-seeing eye which is part of the radar equipment as we know it. This 'eye' can see for five thousand miles ahead and can detect any kind of storm, any object, or any elemental condition that might be approaching in space…"[73]

A similar instrument for surface surveillance is described by Benjamin Creme: "All the space vehicles have listening devices, which may be two or two-and-a-half feet in diameter, and they can be directed from the spaceship itself, which might be – if it is the usual scout ship – 30 feet in diameter. These are sent down and connected electronically with the mothership, and they have instruments which can read the reports which are fed into very advanced computers. (…) these pick up data on everything: the quality of the air, the thoughts and ideas of individuals, if necessary. They may hover outside a room here and listen to this conversation, for example."[74]

Adamski himself observed them around his residence at Palomar Terraces, and major Hans C. Petersen, his co-worker in Denmark, who called them 'telemeter discs', published photographs that were taken of one in Aalborg, Denmark in January 1963.[75]

On their way back to the Moon, Martian scientists explain to Adamski: "By a special device operated in conjunction with this screen we can detect and differentiate between all form-life, be it mineral, vegetable, animal, fowl, or human and even tell its degree of development. (…)

This is the way we measure intelligence of the inhabitants of the Earth or any of the planets."[76]

Although he presented his experience as a children's story, in the recent reissue of his classic *Ami, Child of the Stars*, which provides many corroborations of the accounts of other contactees, Enrique Barrios also describes such a detection instrument and how it is used to 'measure' people's level of empathy or, for instance, their response to a sighting of a spaceship. Because the Universe is the reflection of a perfect, superior order, he is told, there are mathematical laws that apply in all domains, even in the evolution of the civilisations of the Universe: "When human beings share love, affection and kindness, they radiate a certain kind of energy, a very fine energy, the highest, in fact. It can be measured by instruments like those we have… because love is a force, a vibration that penetrates the whole Universe; it is what enables the Universe to exist…"[77]

George Adamski's Martian hosts confirm that all their ships have such an instrument, "and should they come into a strange place, this is the way they would observe it first before landing. They would definitely know from this altitude whether the people they were about to meet would be friendly towards them or not. Beneath the ship is a little apparatus, with a receiver in the ship. This they now turn on and through it we can hear people talking and what they are saying." One of the Martian scientists explains: "It matters not what the language, we would study the vibration of their sound. For the sound of everything carries its own vibration in different pitches for different meanings."[78]

The Technology of Light, says Benjamin Creme, "is no

simple technology which once we have, we have. This technology will eventually give us control over the forces of the universe."[79] Echoing what Howard Menger was told, the Space Brothers, according to Creme, "will put it at our disposal, as soon as we renounce war forever, showing that we are able to live together in peace, with justice, sharing, and right relationship. Then we will know that They are our brothers indeed."[80]

This seems to be corroborated by the visitor who contacted Dino Kraspedon: "We would prefer to see you as horsemen of space, holding the reins of a fiery chariot, or as intrepid sailors braving the turbulent seas of the Cosmos. Maybe men would then come to understand the grandeur of the works of God who dispensed riches abundantly in every corner of the Universe, and see there is no need to fight for land and *lebensraum* [German: 'living space']. To fight for these things shows an ignorance of the greatness of the Universe...

"Maybe men would also cease destroying one another in warfare over some wretched oil wells which are no more important than holes in the ground. If they need energy, space sends it to them from all directions by means of cosmic rays. (...) If they would cease making war, and live like rational beings, we would show them how to harness energy, be it atomic, solar, magnetic or cosmic energy. If they learn to be peace-loving and merciful, the elder brothers in the solar system will show them how to turn this Earth into a Garden of Eden."[81]

In a compilation of his teachings through some associates in London, Maitreya the World Teacher is said to inspire some scientists who are working with the Technology of

Light: "Eventually fewer hospitals will be needed; computers will be adapted to this new technology; medicines we use will become defunct; a new mode of transport will develop, using the magnetic field [of the planet] and the Seven Rays. [In the wisdom teaching the Seven Rays are the seven basic streams of energy that pervade our solar system, our planet and all that lives and moves within its orbit; GA.]

"Mind is like a computer; it needs someone to use it. How do you use the mind? Through the agency of consciousness. Consciousness is governed by light. The moment the relationship is articulated between mind-consciousness-light, you will see big changes. The rays, the colours, will trigger mind and spirit. Then, even a child of five or six will be able to interpret at a deeper level the structure of the atom. This type of civilization already exists on other planets, where this 'science of light' is developed. The aeroplanes and ships we know now will be things of the past. There will be no need of oil or petrol."[82]

At the moment a debate is going on in scientific circles whether or not machines or computers will ever have consciousness. Artificial Intelligence (AI) researchers who believe that consciousness is a product of brain activity seem to be convinced that once our brain functions are fully understood, it will be possible to program them into a computer. This, of course, ignores the fact that the brain itself is merely the physical plane counterpart of the mind as the interpreter between the three-dimensional world of our sensory experiences and our 'self', commonly referred to as the 'soul' – the seat of our individual consciousness. As an increasing body of research and experiences shows, the brain can be dead, leaving the physical body incapacitated, while the soul continues to consciously experience what is going on around it.

These, as yet esoteric, facts notwithstanding, in July 2019 Microsoft announced a $1 billion investment to achieve Artificial General Intelligence (AGI) for machines with the capacity to learn tasks the way human beings do, even though that is "a holy grail of AI that still remains (and may always remain) out of reach".[83] Not least, according to Norwegian neuroscientist and winner of the 2014 Nobel Prize Edvard Moser, because "To simulate the brain, or a part of the brain, one has to start with some hypothesis about how it works." And thus far neuroscientists are largely in the dark about how our brains operate. This is the reason why the European multi-billion-dollar Human Brain Project failed. A commentator incisively points out: "Forget the 'hard problem' of consciousness: We don't even understand the code that neurons use to communicate, except in rare cases. The way that many neurotransmitters affect the activity of neural networks is largely unknown. (…) Many of the specifics of how we form and recall memories are not, for the most part, understood. We don't really know how people make predictions or imagine the future. Emotions remain mysterious. We don't even know why we sleep – in other words, how we spend about a third of our lives is an enigma."[84]

Nevertheless, according to Maitreya the World Teacher, "Scientists have already been taught by the Space Brothers how to use the technology of light to dissolve drug habits by targeting specific areas of the brain. These methods are already being used in hospitals in China and Hong Kong where drug addiction has become a massive problem."[85]

He also says that the Space Brothers have been teaching scientists how to produce energy through light, and how to use this technology to transmit objects from one part of the world to another. "This technology – using colour, sound and vibration – is the science of the 21st century." And in November 1989 he said: "Many experiments with the technology of light are being undertaken in Russia. This is because the Space Brothers set up a community in that country 15 years ago. Russia was chosen because it was such as a closed society, where commercial pressures did not affect scientific research. Today, the Russians are well ahead of the rest of the world in this field. That is why Russia leads the way in calling for arms reductions."[86]

This latter statement was made before the collapse of the Soviet Union and the subsequent immersion of the Russian Federation in the same kind of hyper-capitalism that has engulfed the Western world. Given the fact that this new technology will make other energy sources, including fossil and nuclear fuel, obsolete, it is to be expected that the technology, or perhaps the whole community set up by the Space Brothers, has been guarded from access by those who would use it to advance their own interest over that of the common weal until such time when humanity comes to its senses. Speaking of extraterrestrial help, Maitreya says: "Enlightened beings can control this technology of light but they do not misuse it. In important instances, they have prevented its misuse. Their role is protective and inspirational."[87]

Given that electricity, as the basis for the Technology of Light, produces the cohesion in all forms, it will also be used to create new organs: "Scientists in Russia and America are already experimenting with genetic engineering and this will develop until the time will come when they can transmit genetic information into a sick organ to rejuvenate it without the use of surgery."[88]

As a side note, Benjamin Creme emphasized, "there is genetic engineering and genetic engineering. Much greater care, and much more extensive experiment, is needed before results, in the form of modified crops and animals, should be offered to the public."[89]

Without any evidence this may all sound like wishful thinking, but corroborations may be found in the accounts of several contactees. In his first book in 1978, based on his experiences since the 1950s, Pierre Monnet writes about his extraterrestrial contacts as possessing the knowledge of universal laws which govern the process of spontaneous cell regeneration, which was used on himself: "They can indeed provide our world with very simple means of curing all illnesses, even the most serious." However, "Only man who has developed within himself the law of Love can acquire Knowledge. As paradoxical as it may seem to a Cartesian scientist, the process of spontaneous cellular regeneration giving man the immortality of body and soul, can only manifest itself when he is Love. Only Love is regenerative; for he is Life, eternal, perpetual Life."[90]

Supporting evidence for the claim of extraterrestrials healing humans comes from Ernesto de la Fuente Gandarillas, a civil engineer and former director of technical services for Chilean television. After having been diagnosed with lung cancer in the early 1980s he was offered treatment in a community of extraterrestrials on Isla Friendship in southern Chile, from where he returned fully healed.[91]

And in a lecture in May 1963 George Adamski claimed he had already been shown a machine in the eastern USA in 1955 that would make surgery unnecessary, and where he seems to refer to aspects of electricity: "The molecules that make up the human body are negative and positive, like shoelacing. What happens here? The ray [from the machine] separates (…) the positive from the negative and, when they are separated, there is space, there is nothing there anymore because they suspend the solids of our body. Then they get in there, and replace that bone and put another ray on and harden it (…) all together while you are watching it. No pain, no blood, no cutting. Soon, the molecules get together and close in and no mark is showing."[92]

Benjamin Creme's Master revealed a little more of how the technology of light will be implemented: "The energy of light, direct from the sun, will flow into and from containers of various size while the Power of Shape will determine the nature of the energy needed and stored."[93] Thus, the shape and the size of the containers will determine if the energy stored in them is suited for domestic or industrial use, or for transportation purposes.

An episode in the account of Italian contactee Bruno Sammaciccia in the book *Mass Contacts* seems to refer to such a 'container' in an extraterrestrial underground base near Forlimpopoli, Italy. It looked very different to any electronic device and contained an energy load that would last for a year: "…inside there was a screen, and over it a wide light was moving, without making any noise; it was a light, a dark green one, but it was as if there was some matter in it, maybe one could even touch it. It was like a boiling broth." Mr Sammaciccia's extraterrestrial host explained: "This is energy at its initial state. It may be transformed into solid energy, or into even more subtle energy. And this depends on this small instrument nearby."[94]

As pointed out above, in *Pioneers of Space* George Adamski describes several instances where all the energy needs of our neighbours in space seem to be fulfilled by a technology that runs on light. For instance, on a visit to a "big city" on the Moon, the terrestrial explorers passed by a bakery. Adamski noticed: "They didn't use fire to bake [their bread]. They seemed to utilize light as energy, for dough was placed in a case of glass before our eyes and we watched it bake. It took about three minutes."[95] And about the Martian cruise ship that took them from the Moon to Venus, they are told: "This ship is actually operating by the energy of light, which is the only fuel it uses for its propulsion." Asked if they used any other fuel, their hosts replied: "Never. This is the only energy we use."[96] In fact, he wrote: "Light seems to be the power used for all purposes, illumination, heating, refrigeration, motivation, or energy of any sort, in the home and in business."[97] Likewise, "The whole utility system of the planet [Venus] is operated by light energy…"[98]

With the US Navy finally admitting, in July 2020, that it has video recordings of flying objects which it had to classify as "unidentified", we may expect that the day is approaching when some branch of the government or military will acknowledge it salvaged wreckage from crashed 'flying saucers' in the 1940s.

"After looking into this, I came to the conclusion that there were reports – some were substantive, some not so substantive – that there were actual materials that the government and the private sector had in their possession," says former US Senate Majority leader Harry Reid, who also introduced the bill that funded the AATIP program that ran from 2007-2012. Eric W. Davis, an astrophysicist who worked as a subcontractor and then a consultant for the Pentagon UFO program since 2007, said that, in some cases, examination of the materials had so far failed to determine their source and led him to conclude: "We couldn't make it ourselves." Mr. Davis, who now works for defence contractor Aerospace Corporation, said he gave a classified briefing to a Defense Department agency as recently as March 2020 about retrievals from "off-world vehicles not made on this earth."[99]

The crashes in the late 1940s happened, says Benjamin Creme, as sacrifices made by the space visitors to help humanity accept the reality of the other planets in our system being inhabited by intelligent, benevolent beings. The most famous crash occurred outside the town of Roswell, New Mexico late June 1947,with other crashes documented in the US as well, and several in other countries, including one near Kapustin Yar, the Russian equivalent of Area 51, about 100 kilometres east of Volgograd, in the Soviet Union in 1948.[100]

In addition to the bodies and a surviving crew member – in both the US and the Soviet crashes – technology was retrieved from the wreckage which apparently accelerated advances such as the transistor and touchscreen technology. George Adamski was well aware that the technology used by the UFOs has been much sought after by our governments and military. In a letter to Alberto Perego, the Italian diplomat who was a contactee himself and his main contact in Italy, Adamski wrote: "If a state on Earth could control this energy, it could actually control the world through the dominion of air. I can assure you, however, that these devices are not of an earthly nature. I don't doubt that there are countries where experiments are being carried out to try to build similar devices. Indeed, I am aware that several industrial groups are trying to discover the means to control this energy. Some may have come very close to the goal. But as far as I know, no one has yet come to the full discovery."[101]

Evidence for the claims of retrieved technology was found in the documents related to the Majestic-12 Special Studies Group that were unofficially released (leaked) in the 1990s as part of the US government's efforts towards managed disclosure. These documents are said to have been a briefing for President Eisenhower and although disputed – as controversial materials often are – in his book *Managing Magic: The Government's UFO Disclosure Plan* (2017) researcher Grant Cameron documents that the MJ-12 papers, as they are known, are indeed authentic.[102] In a private conversation Benjamin Creme confirmed that most of the MJ-12 documents were authentic, although some fakes were included[103], probably for reasons of plausible deniability.

Since the crashes and the cover-ups, back-engineering seems to have yielded increasingly advanced propulsion and craft. In June 2019 *The Drive* website reported that the United States Secretary of the Navy is listed as the assignee on several aviation technologies patented by aerospace engineer Salvatore Cezar Pais, claiming that "the U.S. and China are actively developing radical new craft that seem eerily similar to UFOs reported by Navy pilots." (Indeed, in an interview with *Newsweek* magazine, Harry Reid "dropped major hints that he knows potential adversaries, Russia and China, have carried out their own military studies to figure out how UFOs work and how to build their own".[104]) When the patent application was initially rejected for being based on non-existent technologies, Chief Technical Officer of the Naval Aviation Enterprise Dr James Sheehy personally vouched for it in a letter to the US patent office, claiming the Chinese are already developing similar capabilities. The website reports: "The hybrid aerospace-underwater craft in Pais' patent, meanwhile, is described as being capable of incredible feats of speed and maneuverability and can fly equally well in air, water, or space without leaving a heat signature. This is possible, Pais claims in the patent, because the craft is able to 'engineer the fabric of our reality at the most fundamental level' by exploiting the laws of physics."[105]

However, the spacecraft that the visitors use require more than just advanced technology. According to Grant Cameron, "in 1947, on day one when this thing starts [with the US Air Force retrieving the wreckage and casualties from a crashed saucer], they realize there's this mental aspect, and realize that we would learn later, is that's how you fly the craft. You use your mind."[106] This was also witnessed by Adamski, who said that "while there were all types of instruments in the ship they were not ordinary ones since they were subject to the pilots' own consciousness."[107] Others who say they were invited on board by extraterrestrial visitors also reported or hinted that the craft are operated by the minds of the pilots interfacing with the craft's advanced technology.

If the technology of the extraterrestrial space craft is operated by mental powers or consciousness there is no reason to assume that any such technology is currently in the hands of the US, Russia, China or India, to name the most obvious contenders. As I pointed out before, this has been confirmed in the past by several contactees. And as recently as 2008 Apollo 14 astronaut Dr Edgar Mitchell stated that any back-engineering "is not nearly as sophisticated yet as what the apparent visitors have".[108] Dr Mitchell was also the founder in 1973 of the Institute of Noetic Sciences, one of the earliest organisations to research consciousness.

In the current world of globalized cut-throat competition for natural resources, consumer dollars and the economic power that these yield, even the suspicion that one nation has a technological advantage over another could easily lead to reckless attempts to outsmart the competition and descend into nuclear conflict, even if unintended. This is so obvious that in Enrique Barrios' children's book about extraterrestrial visitors we read: "When the scientific level overwhelms the level of love in a world, that world self-destructs. There is a mathematical relationship…"[109]

In June 2019, Robert Wood, who has done invaluable work as the chief researcher of the Majestic-12 documents, gave a talk for the Society for Scientific Exploration, titled 'The secret relations between UFOs and Consciousness'. Here he showed an official document from the 1950s which specifically states that, in his words, "deceptive methods are to be used to confuse the public about the reality of the recovery of downed UFOs." This confirms that governments and the military colluded in a campaign to spread disinformation about the craft and visitors from space.

This campaign was so successful that not only did the notion of UFOs and space visitors become a laughing stock, it also prepared the field for ever wilder speculations. Although a scientist himself and an otherwise respected researcher, in the same lecture Dr Wood claims what, astonishingly, is accepted as fact in conspiracy-prone circles: "After the Nazis came into power they had explored the Antarctic caverns and signed a treaty with an evil alien species who would help them try to conquer the world. The aliens donated some completed craft to the Germans who [inaudible] trying to manufacture them for the war."[110] And at this point in his talk he shows a diagram of a saucer-type craft labelled 'Haunebu II'.

That this claim lacks any logic or common sense should be clear from what follows. In 1955 George Adamski acknowledged: "We have been making fun at Flying Saucers, yet it is amazing how much we already learned from them. (…) They have not only awakened our minds to the potentials ahead of us, not only to the potentials of other human beings like ourselves, living on another [planetary] body there, floating through space as we are … but they taught us *what* to look for in order to do the same thing. They have taught us much more than has been written down, or given the public in general."

At another lecture in Detroit that month, he told his audience how he met Hermann Oberth, the father of rocket science and astronautics, in Buffalo, New York: "There were a lot of things we talked about that I can't go out and speak of publicly." To experience these, he says, we would have to "let something else guide you and *not* the traditional and conventional standards in which man has grown... and which have gone far from nature."[111]

In a talk at the Detroit Institute of Arts on 20 September that year, George Adamski showed his audience a diagram from Germany, saying: "The Germans were already working on a Flying Saucer. It tells you all about it here... there are diagrams of all kinds. It was mostly propelled by jet propulsion, as you can see."[112]

The fact that a regime ambitious to rule the world attempts to develop the means to help them do so, comes as no surprise. But given the various claims that the Nazi's actually managed to develop antigravity propulsion and subsequently set up camp in a secret base on Antarctica – or Mars, depending on your conspiracy theory of choice – they have apparently been very reluctant to employ their technological advantage against the world. Unless the rest of us missed the memo about the Nazis ruling the world since 1945.

Besides, in a meeting with close associates who were asked to publish his forecasts between 1988 and 1993, the World Teacher said in October 1989 about technology that is already in the hands of the Americans and the Russians, that the Space Brothers are watching: "Scientists will not be allowed to use technology to interfere with the laws of creation. Their technology might have to be destroyed because it could even interfere with the evolution of the [Space] Brothers themselves."[113]

On his first meeting with a being from Saturn during

his out-of-body visit to the Moon, George Adamski was informed: "Right now the Earth man is so insane that he is no longer the man the Creator meant him to be. He hasn't room enough to fight his own world, thereby destroying it bit by bit, along with the labors of its people, but he is even now trying to get high enough above the Earth to destroy it faster. In this he is not going to succeed, not at least to the degree he expects." He added that although we may acknowledge in words that we are children of the Creator, "By your deeds you deny it, for you destroy your brothers."[114]

Nevertheless, the rumours of extraterrestrial technologies being in the hands of Earthly powers abound, clearly aimed at feeding the fear for an alien invasion that has been nourished through the disinformation efforts since the mid-1950s, and have found a loyal following among a segment of the population that thrives on thrills at the prospect of an alien scapegoat for humanity's own failings. Not to mention the segment of the population that is ruled entirely by fear of the extraterrestrial presence. That a certain number of people are susceptible to such fabrications is to be expected – and is reflected in the vocal groups who fear that government measures to stem the spread of Covid-19 are designed to enslave the people.

The situation becomes more problematic when 'big' names step forward to promote hypotheses or 'evidence' of Nazi descendants who have their hands on alien technology, such as Robert Wood, former Canadian Defence minister Paul Hellyer, or rock musician Tom DeLonge. The latter is the former frontman of the band Blink-182 and since 2016 presents himself as the chosen liaison between various government or military insiders and the public, entrusted with the task of gradually disclosing what the government knows about extraterrestrial visitors. However, in *Managing Magic* Grant Cameron meticulously documents how Mr DeLonge's position is more accurately that of a pawn in the shady agencies' chess game of going about UFO disclosure. After decades of blatant lies and furtive efforts to actively mislead the public, the authorities can no longer afford to just step forward and announce the truth about the extraterrestrial visitors. Moreover, the dark forces of materiality behind the undemocratic workings of some governments are still invested in maintaining their hold on their populations by nourishing thoughtforms of impending catastrophe.

According to Benjamin Creme in 2003, similar forces that worked through the Nazis in Germany, the militarists in Japan and fascist Italy during WWII, are now working – not from Antarctica, but through the Pentagon in the US, the Zionists in Israel and a group of former Eastern Bloc countries in Europe, albeit to a lesser extent than in 1933-1945. As an example, Creme said, "This war in Iraq and the general thrust of the ambitions of the US Republican government and of the Israeli government's oppression of the Palestinian people, are part of the same energetic outflow."[115] A succinct history of the route that fascism took from the defeated Nazis to the Pentagon, for one, rather than the caverns of Antarctica, was outlined by Daniel Sheehan, a constitutional trial attorney and social justice and UFO disclosure advocate, in an interview in 2018.[116] This confirms that the threat does not come from outside the Earth, but from within humanity's own ranks.

Of course, to divert our attention from their own dark

motives, the Pentagon, or their shady agencies, need someone to frame, and after many decades of misinformation and disinformation, what better target than those elusive 'aliens'? All the pieces have been carefully put in place for 'full disclosure', including the involvement of 'insiders' and 'experts' who have seen 'documents'. And as Grant Cameron summarizes: "The DeLonge material described ... that there are evil aliens here, they are up to no good, and 'money and resources will be needed by the military industrial complex for the battle against the alien invasion'."[117] Mr Cameron also references the work of longtime disclosure advocate Dr Steven Greer who "accurately points out [that] the evil alien theory 'coopts the progressive community' of UFO researchers into a dualistic worldview of good blond human aliens versus evil grey and reptilian races, which seems to be an extension of many humans' racial and ethnic tensions with one another. We are therefore projecting on to outer space our existing frailties."[118]

Another warning comes from Dr Carol Rosin, who was the spokesperson for Dr Werner von Braun during the last years of his life, and the founder of the Institute for Security & Cooperation in Outer Space that has been promoting the Treaty on the Prevention of the Placement of Weapons in Outer Space.[119] Speaking about Von Braun, who went from being the leading rocket scientist in Nazi Germany to becoming a decorated pioneer of space technology in the US, Dr Rosin says: "What was most interesting to me was a repetitive sentence that he said to me over and over again during the approximately four years that I had the opportunity to work with him. He said the strategy that was being used to educate the public and decision makers was to use scare tactics." According to Von Braun the Russians would be the first 'enemy', then the 'terrorists', followed by 'asteroids', and "the last card is the alien card. We are going to have to build space-based weapons against aliens and all of it is a lie."[120] It seems that those who stand to lose most from a humanity that realises its oneness are now playing this card through an unwitting Tom DeLonge and his band of partly (or mis-)informed 'insiders'.

The fact that the 1950s contactees, without exception, reported on the friendly, or rather, brotherly nature of their contact with visitors from space doesn't mean there are no planets that are less evolved and whose inhabitants may have their own more selfish agendas. The Master from Venus, for instance, told Adamski: "In the vastness of space there are innumerable bodies which you on Earth call planets. These vary in size, as do all forms, but they are very much like your own world and ours, and most of them are peopled and governed by forms like yourselves and like us. While some are just reaching a point where they are capable of supporting such forms as ours, others have not yet reached that stage of development in their growth."[121]

In this regard, Benjamin Creme said: "There are various tales in magazines and newspapers of people being taken up, experimented on, and things being inserted under their skin and so on. All of this is totally untrue. There is not a single instance of such happenings. All of these stories are the result either of the fevered astral imagination of people who *want* to feel these things and do so in an astral sense, which they then describe to others and so build up a climate; or work of certain negative forces in the world whose aim is to keep from the public the reality of the extra-terrestrial connection of this planet."[122]

Based on his contacts Pierre Monnet said: "On each planet there are different stages in the evolution of [human] beings. It goes from the most loving to the most aggressive, depending on the degree of vibration reached by each individual; it also goes from the most active to the most inactive and from the most conscious to the most unconscious."[123] As the current polarisation shows, we have plenty of both extremes on our own planet.

But keeping in mind that everything in cosmos evolves according to a Plan, and given the volatile psychology of Earth humanity at this point in our evolution, three independent sources have stated that the more evolved space brothers have placed a ring of protection around the Earth, so that humanity is free to address its own materialist tendencies without being negatively influenced from outside. Wilbert Smith referred to it as a "cosmic police force"[124], Benjamin Creme called it a "ring of light"[125], and scientist Michael Wolf spoke of an "alien barrier" that bars hostile ETs from coming here.[126]

What's more, they are actively involved in keeping the planet habitable in the face of our own worst efforts to the contrary. In the words of Benjamin Creme: "One of the main factors in maintaining our eco-system is our Space Brothers: we owe them an enormous debt." He says: "Within karmic limits they mop up as much radiation and pollution as possible. They also go down into the oceans and neutralize waste which we have dumped there and which otherwise would kill off marine life and further poison the planet."[127] Indeed, his Master said: "Life on this planet would be utter misery were it not for the help of our

Space Brothers who neutralize this pollution and render it harmless within karmic limits. Fleets of Their space ships, using implosion devices, do this on daily basis."[128]

When a bright blue flash was filmed near Tokyo during the heavy earthquake there in 2011, and later during an earthquake in Chiapas State, Mexico, in 2017, Benjamin Creme explained that the Space Brothers had adapted the implosion technology with which they mitigate nuclear radiation, to mitigate the force of earthquakes, causing the bright blue flashes. In confirmation of statements from earlier contactees, Pierre Monnet, too, said that the space visitors "are doing their utmost to reduce the intensity of our earthquakes and limit the damage."[129]

Exactly how consciousness may interface with and control technology is beyond my understanding of either, but in the posthumously published *The New Science*, which Wilbert Smith wrote based on his own engineering background, research, and "data obtained from Beings more advanced than we are", he gives an interesting hint: "Just as Area has Length incorporated in it, and Volume has Area incorporated in it, so has the Electric Field the Tempic Field incorporated in it and the Magnetic Field has the Electric Field incorporated in it. Each of these three fields are mutually at right angles to each other. The three fields together are the manifestations of Reality in the Field Fabric as perceived by Awareness. The interrelationships between these various fields manifests to our Awareness as Matter and Energy, and the great variety of these manifestations is well known to us."[130]

The way I understand this is that matter and energy,

on which our current technology is based, may be connected or interrelated with our consciousness in the three-dimensional (tempic) field through the electric and magnetic fields. I assume a proper understanding of how these fields are interrelated and interdependent would make it possible to design technology that may be controlled by consciousness, through the power of thought.

With this in mind, it is interesting to remind ourselves of what Daniel Fry was told about science: "Every civilization in the Universe, no matter where or when it originates, develops primarily through the continuing increase in knowledge and *understanding* which results from the successful pursuit of 'science'.

"The word 'Science' has been defined in your dictionaries as 'the orderly, and intelligently directed, search for truth.' Under this definition, the whole of science may be divided into three principal parts... (1) The physical or material science; (2) The social sciences – relationship between man and his fellow man; and (3) The spiritual science – relationship between man and the great creative power and infinite intelligence which pervades and controls all nature. If any civilization in the Universe is to develop fully and successfully, each of the three branches of science must be pursued with equal effort and diligence." And confirming the primacy of consciousness, "The Spiritual and Social sciences, however, must come first. There can be no dependable development of a material science until you have first built a firm foundation of spiritual and social science."[131]

Benjamin Creme makes a comparable tripartition: "There is concrete science or technology. There is the science of the higher mind – philosophical, theoretical and abstract – the science of Einstein, for example. There is also the science of the psyche or white magic, which the Masters use. It is the same science as the others, but is intangible, although you can see its results."[132] Both seem to point to the distinction that needs to be made between the knowledge of the lower, concrete mind; and the intuition or wisdom of the higher, abstract mind. Likewise, it seems that only by recognizing these various aspects of science, dealing as they do with the various aspects of reality, will it be possible to develop – and indeed use – the technology to navigate space in the way that the space visitors are capable of.

About his search for the science behind the UFO phenomenon Bruce Cathie muses: "Here we are, searching frantically for something that is right in front of our eyes, always has been, and always will be; the simple truth that *we* are part of the Creator; this is the meaning behind true phrases such as: 'the equality of all men', 'all men are brothers', and so on, for as cells among the uncountable billions of cells, we are all aspects of God, God is Us, We are Light. When we know how to use the laws of harmony which permeate the whole of Creation, we shall be free to create that which is now unattainable. *We* hold the keys to our own salvation... Conscience, the laws of cause and effect, the teachings of superior brothers from space – these are the guides."[133]

Likewise, a Martian scientist told Dr Johnston in *Pioneers of Space*: "Once men of science have learnt all of this, more than half of their present knowledge will have to be cast aside. That part will be the complicated part and it must be replaced by simplicity. Another thing that all men of science must do on Earth … is to become science, by living science and not merely for honors and rewards. All sciences should be coordinated to be as one, not divided as they are."[134]

As if emphasizing the need for a systems approach or a post-material science, he continues: "To be a full fledged scientist one must acknowledge all phases of manifestation, be they labelled spiritual or material, so long as they are natural. (…) We have found by working together as one, our findings are almost absolute, for when any one branch of science discovers something or comes upon a problem, he consults all other branches to solve it, while scientists on Earth, astronomers as they are named, seldom consult other branches which are considered lower than theirs. Do not forget that this vastness is one system and takes in all branches of science."[135]

In the conclusion to the Galileo Commission Report Dr Harald Walach concurs: "There is no empirical or theoretical ground for holding on to the current concept [of science] except an old fashioned and barely understood ideology [that considers matter fundamental, and consciousness an effect]. We have argued that it should be set aside and broadened out into a spirituality informed science. And we are excited about new options that will arise from it."[136]

According to Benjamin Creme's Master, such a new approach to science will lead to a complete readjustment in our attitude to reality: "The new science will show humanity that all is One, that each fragmented part of which we are aware is intimately connected with all others, that that relationship is governed by certain laws, mathematically determined, and that within each fragment is the potential of the Whole. This new knowledge will transform men's experience of the world and of each other and confirm for them the truth that God and man are One."

He also says that our hidden psychic powers, of which Vera Stanley Alder said technology is the externalized aspect, "will unfold naturally and the vast potential of the human mind will conquer space and time and control the energies of the universe itself."[137]

References

1 Benjamin Creme (ed.), 'Signs of the time'. *Share International* magazine, Vol.34, No.3, April 2015, p.12
2 Gretchen McCartney, 'Mystery Solved: Bright Areas on Ceres Come From Salty Water Below'. NASA's Jet Propulsion Laboratory, 10 August 2020. See: <www.jpl.nasa.gov/news/news.php?feature=7722>
3 Agence France Press, 'Planet Ceres is an 'ocean world' with sea water beneath its surface, mission finds'. *The Guardian*, 10 August 2020. See: <www.theguardian.com/science/2020/aug/10/planet-ceres-ocean-world-sea-water-beneath-surface>
4 Gerard Aartsen (2015), *Priorities for a Planet in Transition*, p.23
5 Bruce Cathie (1968), *Harmonic 33*, p.151
6 Sean Walker, 'The Mysterious Martian "Plumes" '. *Sky&Telescope*, 24 February 2015. See: <skyandtelescope.org/astronomy-news/transient-martian-phenomena-022420155/>
7 European Space Agency, 'Mysteriously Long, Thin Cloud Returns on Mars – Not Linked to Volcanic Activity'. *SciTechDaily*, 31 July 2020. See: <scitechdaily.com/mysteriously-long-thin-cloud-returns-on-mars-not-linked-to-volcanic-activity/amp/>
8 Sara Seager, 'Phosphine Gas Detected on Venus'. Nature.com, 14 September 2020. See: <astronomycommunity.nature.com/posts/phosphine-gas-detected-on-venus?badge_id=280-nature-communications>
9 Vera Stanley Alder (1940), *The Fifth Dimension and The Future of Mankind*, pp.92-96
10 Interview with Pierre Monnet, *L'Ère Nouvelle*, January 2006. See: <pointdereference.free.fr/m/www.erenouvelle.com/PORTCO-7.HTM>
11 Enrique Barrios (1989), *Ami, Child of the Stars*, p.84
12 Daniel Fry (1954), *The White Sands Incident*, pp.77-78
13 Alder (1940), op cit, pp.93-94
14 'Psychic Abilities'. Talk by Russell Targ for SUE Speaks, Mighty Companions, 2013. See: <www.youtube.com/watch?v=hBl0cwyn5GY

15 Harald Walach (2019), *Beyond a Materialist Worldview. Towards an Expanded Science*, p.58

16 Jim Armitage, 'Even worse than Foxconn: Apple rocked by child labour claims'. *The Independent*, 30 July 2013. See <www.independent.co.uk/lifestyle/gadgets-and-tech/even-worse-than-foxconn-apple-rocked-by-child-labour-claims-8736504.html>

17 Michael Sainato, ' "I'm not a robot": Amazon workers condemn unsafe grueling conditions at warehouse'. *The Guardian*, 5 February 2020. See: <www.theguardian.com/technology/2020/feb/05/amazon-workers-protest-unsafe-grueling-conditions-warehouse>

18 Mike Mwenda, 'Tech giants sued over child labour, deaths and injuries in cobalt mining in the DRC'. *Lifegate*, 29 January 2020. See: <www.lifegate.com/cobalt-mining-congo-tech-lawsuit>

19 Alan Rusbridger, 'Amid our fear we're rediscovering utopian hopes of a connected world'. *The Observer*, 29 March 2020. See: <www.theguardian.com/commentisfree/2020/mar/29/coronavirus-fears-rediscover-utopian-hopes-connected-world>

20 Barrios (1989), op cit, pp.37-38

21 Michael Hesemann (1996), *The Cosmic Connection. Worldwide Crop Formations and ET Contacts*, p.7

22 Alick Bartholomew (ed.; 1991), *Crop Circles – Harbingers of World Change*, p.13

23 Lucy Pringle (1999), *Crop Circles. The Greatest Mystery of Modern Times*, p.101

24 Horace R. Drew (2012), 'Hard factual evidence for three paranormal crop circles…', Appendix 4. 'Further evidence of paranormal crop pictures: the effects of unknown energies on plant morphologies in oilseed rape or wheat'. See <www.cropcircleconnector.com/anasazi/time2012a.html>

25 Cathie (1968), op cit, p.28

26 Ibidem, p.12

27 Creme (1993), *Maitreya's Mission*, Vol. Two, p.210

28 Creme (2010), *The Gathering of the Forces of Light – UFOs and their Spiritual Mission*, p.190

29 Benjamin Creme's Master, 'Invisible peril', June 2006. In: Creme (ed.; 2017), *A Master speaks*, Vol. Two, p.62

30 Cathie (1968), op cit., p.202

31 George Adamski (1949), *Pioneers of Space*, pp.84-85

32 Creme's Master, June 2006, op cit. In: Creme (ed.; 2017), op cit, p.61

33 Bailey (1925), *A Treatise on Cosmic Fire*, p.910

34 'Scientists Can Now Recycle Water, Air, Fuel, Making Deep Space Travel Possible'. Sputnik, 16 July 2018. See: <www.spacedaily.com/reports/Scientists_Can_Now_Recycle_Water_Air_Fuel_Making_Deep_Space_Travel_Possible_999.html>

35 Adamski (1949), op cit, p.75

36 Creme (2001), *The Great Approach – New Light and Life for Humanity*, p.130

37 Ibid., p.227

38 Creme (1997), *Maitreya's Mission*, Vol. Three, p.334

39 Creme (2010), op cit, p.127

40 Ibid., pp.185-86

41 Adamski (1949), op cit, p.197

42 Ibid., pp.120-21

43 Ibid., p.198

44 Patricia Pitchon (1994), 'Interview with Benjamin Creme's Master: Closing Nuclear Reactors and Discovering New Energies'. In: Creme (1997), op cit p.194

45 Howard Menger (1959), *From Outer Space to You*, p.47

46 Fry (1954), op cit, pp.44-45

47 Adamski (1949), op cit, pp.81-82

48 Xiameng Liu et al, 'Power generation from ambient humidity using protein nanowires'. *Nature.com*, 17 February 2020. See: <www.nature.com/articles/s41586-020-2010-9.epdf>

49 Bailey (1936), *Esoteric Psychology*, Vol.I, pp.373-74

50 Adamski (1949), op cit, pp.190-91

51 H.P. Blavatsky (1888), *The Secret Doctrine*, Vol.I, p.554 (6th Adyar ed., Vol.II, p.278)

52 Creme (2010), op cit, p.185

53 Blavatsky (1888), op cit, Vol.I, p.110 (6th Adyar ed., Vol.I, p.170)

54 Creme (2001), op cit, p.231

55 J.D. Rabbit, 'Circles of light'. *Share International* magazine, Vol.19, No.7, September 2000, p.21

56 Creme (2010), op cit, p.193

57 Creme (ed.), 'Letter to the editor'. *Share International* magazine, Vol. 24, No.8, October 2005, p.25

58 Creme, 'Questions and answers'. *Share International* magazine, Vol. 23, No.6, July/Aug 2004, p.38

59 John P. Bessor, 'Oregon's strange whirlpool of force'. *Fate* magazine, July 1951, pp.24-25

60 Cathie (1972), *Harmonic 695 – The UFO and anti-gravity*, opp. p.80, and pp.114-15

61 'Mystery spots'. See: <www.roadsideamerica.com/story/29062>

62 Creme (ed.), 'Letters to the editor'. *Share International* magazine, Vol. 20, No.1, Jan/Feb 2001, p.43

63 Creme's Master, 'The Blueprint of the future', October 1999. In: Creme (ed.; 2004) *A Master speaks*, p.359

64 Adamski (1949), op cit, p.221

65 Ibid, p.178

66 Ibid, p.221

67 Stefan Denaerde (1982), *Contact from Planet Iarga*, pp.102-03

68 Creme (1997), op cit, p.182

69 Creme's Master, 'Leisure is the key', November 1986. In: Creme (ed., 2004), op cit, p.101

70 Adamski, *Cosmic Bulletin* newsletter, December 1964

71 Adamski (1955), *Inside the Space Ships,* p.60

72 Ibid., p.77

73 Adamski (1949), op cit, p.70

74 Creme (2010), op cit, p.33

75 Hans C. Petersen (ed.), 'Photo of a Telemeter-Disc', *UFO Contact*, February 1963, pp.74-75

76 Adamski (1949), op cit, pp.249-50

77 Barrios (2020), *Omi of the Stars*, p.24

78 Adamski (1949), op cit, p.73

79 Creme (1997), op cit, p.207

80 Creme (2010), op cit, p.24

81 Dino Kraspedon (1959), *My Contact With Flying Saucers*, p.105

82 Creme (ed; 2005), *Maitreya's Teachings. The Laws of Life*, p.226

83 'Microsoft is investing $1 billion in OpenAI to create brain-like machines'. *MIT Technology Review*, 22 July 2019. See:

84 Tim Requarth, 'The Big Problem With "Big Science" Ventures – Like the Human Brain Project'. *Nautilus*, 'Facts So Romantic' blog, 22 April 2015. See: <nautil.us/blog/the-big-problem-with-big-science-ventureslike-the-human-brain-project>

85 Creme (ed; 2005), op cit, p.118

86 Ibid., p.232

87 Ibid., p.235

88 Ibid., p.229

89 Creme (2001), op cit, p.269

90 Interview with Pierre Monnet, op cit

91 Aartsen (2011), *Here to Help: UFOs and the Space Brothers*, pp.76-78

92 Petersen [n.d.], *Report from Europe 1963*, p.116

93 Creme's Master, 'The cities of tomorrow', April 2008. In: Creme (ed.; 2017) *A Master Speaks*, Vol Two, p.100

94 Stefano Breccia (2009), *Mass Contacts*, pp.190-91

95 Adamski (1949), op cit, p.78

96 Ibid, p.120

97 Ibid, p.163

98 Ibid, p.219

99 Ralph Blumenthal and Leslie Keane (2020), 'No Longer in Shadows, Pentagon's U.F.O. Unit Will Make Some Findings Public'. *New York Times*, 23 July 2020. See: <www.nytimes.com/2020/07/23/us/politics/pentagon-ufo-harry-reid-navy.html>

100 Steve Johnson, 'Russian Roswell'. *UFO Magazine*, [issue, date unknown] 2006. See: <www.rolfwaeber.com/mystery/russian-roswell/index.html> John Alon Walz (dir.), *UFO Files: Russian Roswell*, History Channel, 31 October 2005. See: <www.youtube.com/watch?v=PlqtnkX1kZA>

101 Adamski, letter to Alberto Perego, 20 April 1956, as reproduced in Alberto Perego (1963), *L'aviazione di altri pianeti opera tra noi*, p.539. Author's translation from Italian.

102 Grant Cameron (2017), *Managing Magic: The Government's UFO Disclosure Plan*, 'Disclosure Efforts 1980-1990', pp.92-124

103 Creme, in a private conversation with the author, September 2011

104 Andrew Whalen, 'U.S. in UFO Race With China, Russia, Former Senate Majority Leader Suggests'. *Newsweek* magazine, 1 March 2019. See: <www. newsweek.com/ufo-2019-harry-reid-china-russia-senate-unexplained-aerial-phenomena-1349256>

105 Brett Tingley and Tyler Rogoway, 'Docs show Navy Got 'UFO' Patent Granted by Warning Of Similar Chinese Tech Advances'. *The Drive*, 28 June 2019. See <www.thedrive.com/the-war-zone/28729/docs-show-navy-got-ufo-patent-granted-by-warning-of-similar-chinese-tech-advances>

106 Alex Tsarakis (2016), 'UFO researcher Grant Cameron has uncovered 100s of previously classified UFO documents pointing toward a UFO/consciousness link'. *Skeptiko*, 11 August 2016. See: <skeptiko.com/grant-cameron-ufo-consciousness-link-324/>

107 Adamski (1962), *Special Report – My Trip to Saturn*, Part 2, p.1

108 Interview with Dr Edgar Mitchell in Nick Margerrison, *The Night Before*, Kerrang Radio!, 23 July 2008

109 Barrios (1989), op cit, p.20

110 Bob Wood (2019), 'The secret relations between UFOs and Consciousness'. Talk for the Society for Scientific Exploration, June 2019. See: <www.scientificexploration.org/videos/the-secret-relations-between-ufos-and-consciousness>

111 Adamski (1974), *Many Mansions*, p.15

112 Adamski (1956), *World of Tomorrow*, p.10

113 Creme (ed; 2005), op cit, p.231

114 Adamski (1949), op cit, p.125

115 Creme (2006), *The Art of Living – Living Within the Laws of Life*, p.175

116 Ashley Dickout, 'Daniel Sheehan – The Deep State and Alien Disclosure'. Nsuho, 5 July 2018. See: <www.youtube.com/watch?v=z9E0HcbchRY>

117 Cameron (2017), op cit, p.209

118 Ibid, p.208

119 See: <peaceinspace.com>

120 Carol Rosin, as quoted in Cameron (2017), op cit, pp.210-211

121 Adamski (1955), op cit, p.86

122 Creme (2001), op cit, pp.130-31

123 Interview with Pierre Monnet, op cit

124 Wilbert B. Smith (1969), *The Boys from Topside*, p.45

125 Creme (1979), *The Reappearance of the Christ and the Masters of Wisdom*, p.206

126 Paola Harris (2008), *Connecting the Dots. Making Sense of the UFO Phenomenon*, p.159

127 Creme (2001), op cit, p.130

128 Pitchon (1994), 'Interview with Benjamin Creme's Master'. In: Creme (1997), op cit, p.192

129 Interview with Pierre Monnet, op cit

130 Smith (1964), *The New Science*, 1978 reprint, p.19

131 Fry (1954), *[A Report by Alan] To Men of Earth*. In: Fry (1966) *The White Sands Incident*, pp.75-76

132 Creme (2012), *Unity in Diversity – The Way Forward for Humanity*, p.48

133 Cathie (1972), op cit, p.162

134 Ibid., p.237

135 Adamski (1949), op cit, p.241

136 Walach (2019), op cit, p.88

137 Creme's Master, 'The new civilization', August 1982. In: Creme (ed., 2004), op cit, pp.13-14

"Time is (...) the element action creates in its path from the formless to the formed." –George Adamski, teacher and contactee

5. DISCLOSURE:
THE TIME IN REALITY IS NOW

The UFO phenomenon won't be understood until we have disclosure, according to UFO disclosure activist and executive director of the Paradigm Research Group Stephen Bassett in a recent interview[1], and from a strictly physical point of view perhaps he has a point. Approaching it from the broader view of reality, however, UFO researcher Grant Cameron says: "Consciousness is the elephant in the room when it comes to full disclosure, and it is a key component to the mystery. Including the role of consciousness in the disclosure announcement will lead to the collapse of scientific materialism in (...) the mother of all paradigm shifts."[2]

Throughout our investigation we keep coming across the urge to find a new paradigm. It engages the Foundation for Research into Extraterrestrial Experiences, the Laszlo Institute for New Paradigm Research (Ervin Laszlo), New Paradigm College (UFO disclosure advocate and attorney Daniel Sheehan), the Paradigm Research Group (Stephen Bassett), as well as many other UFO researchers. The fact that a 'new paradigm' has become something of a philosopher's stone in the fields of consciousness studies and UFO research, indicates that the human mind is ready for a new level of understanding.

Around the time that the UFO phenomenon had largely been made into a mystery, in 1964, George Adamski already emphasized that the reason for the extraterrestrial presence is more important than their advanced technology: "The time has arrived that the study of what the Brothers have already given is more important than the saucers, if we are to have a better world, or prepare ourselves for the things yet to come... The individual life is more important to understand than all the spacecraft in the cosmos."[3]

In the preceding chapters we have seen how systems science and quantum science have established that the oneness of life and consciousness – already postulated in the wisdom teachings in the late 19th century and in the accounts about the space visitors since the 1950s – is the fundamental reality of existence.

We have also seen how this oneness, in the process of creation and evolution has fragmented into individual bits and pieces, planets and galaxies, species and beings, all with their own characteristics, goals, and interests. How realistic, then, is it to expect that this 'fragmentation' on even our own planet would ever be overcome to create a world of peaceful, right human relations through unity in diversity as proposed in the messages that the contactees were given? Just looking at the headlines from the past century, without even going into the details of cruelty and unnecessary suffering, would send any hope for lasting peace out of the window, many would say.

As with consciousness, evolution, and technology,

though, it really depends on which perspective we decide to take when we look at the state of the world and humanity and the urgent need to acknowledge our oneness. So before rushing to conclusions, let us take a minute and step back to see how the bigger picture unfolds.

The paradigm shift that so many are looking for has already been outlined in the Ageless Wisdom teaching since the work of Mme Blavatsky and holds that consciousness is fundamental to what we perceive as reality. This perception of reality evolves as our awareness of – and identification with – the source of consciousness grows. Eventually, this leads to mastery over the three-dimensional spacetime domain in which our lives currently manifest. This is a continuous journey that we are all engaged in, and the teachings tell us that there are those who have achieved ahead of us and are ready to re-enter our lives to lighten the way for us.

Indeed, as Benjamin Creme has maintained from the start of his mission in the 1970s, it is in this context that the UFOs have been particularly active since the 1950s: "There is a definite relationship, in the sense that all the Hierarchies in this solar system work together, and what we call the U.F.O.'s (the vehicles of the space people, from the higher planets) have a very definite part to play in the building of a spiritual platform for the World Teacher, preparing humanity for this time."[4] He says, "The reality is that this planet, and humanity on this planet, are part of a Brotherhood which embraces the whole of the solar system, and each is closely inter-related. The energy from this planet streams into every other planet, and the energy from every other planet streams into all planets, including this one. It is a close energetic co-relationship. We have to realise this, and that our thoughts and actions create an effect on the aura of this planet which in turn affects every other planet in this system. If we are responding in a certain way, so that the light and energy emitted from this planet is at a relatively low vibration, we are holding up the advancement of the solar system as a whole."[5]

According to the Ageless Wisdom teaching, the history of humanity on this planet goes back 18.5 million years – most of which is unknown to mainstream science or official history. As with planetary evolution, human evolution does not follow a straight upward slope from primitive cavemen to space exploring businessmen in self-driving cars, but takes place in cycles, very much like the seasons in nature.

Every cycle brings us to the same point, but on a higher turn in the upward spiral. This applies not only to the physical aspects of life, such as the human body and our control over our environment and the tools and technology to do so. It equally applies to the evolution of our emotional, mental and psychological nature. So, while the first physical human race, which is usually named after the Lemurian cycle, took millions of years to develop an adequate physical body, the second physical race in the Atlantean cycle took millions of years to develop a responsive emotional apparatus.

The current human race of the Aryan cycle – which has nothing to do with Hitler's appropriation of the name, but derives from Arya, the ancient name for Persia[6] – including its many ethnic varieties, is engaged in developing control of our mental equipment, the human mind. So far, we have only just begun to use the lower, concrete levels of mind, and this has already brought us the incredible technological achievements of the past few centuries. And while these pale in comparison with the technology available during the Atlantean period, the latter wasn't the achievement of mankind, but a 'gift from the Gods' – the Masters of Wisdom of the time, who ruled as philosopher-kings, and probably the Brothers from space.

In this context, it should be clear why genesis stories based on the modern interpretation of Sumerian tablets by Zecharia Sitchin, or in cosmologies such as suggested by Erich von Däniken, present fragments of the actual history of humanity at best and are mostly well off the mark. Likewise, hypotheses about the many (very real) geographical, architectural, paleontological and other anomalies that feature in popular 'ancient aliens' theories remain hopelessly speculative. Here, as in UFO research, we cannot hope to find consistently coherent answers as long as we ignore the wisdom teachings.

An interesting, although tangential reference to previous cycles of human evolution can be found in the travel diary of the 17th century Belgian Jesuit missionary and cartographer Albert d'Orville. During his stay in Lhasa between October and November 1661 he wrote in his diary: "My attention was drawn to something which moved in the sky. At first, I thought it was an unknown kind of bird which lived in this land, until the thing came nearer, and took the form of a double Chinese hat. While it flew silently, as if it were carried by invisible wings, it was certainly a wonder – magic. The thing flew over the town as if it wanted to be admired, flew twice in circles and was then enveloped in fog, and disappeared." When he asked a nearby Lama if he had seen it too, the reply came: "My son, what you have seen was not magic. Beings from other worlds have been traveling the oceans of space for centuries, and brought spiritual light to the first human beings who populated the earth. They condemn all violence and taught the people to love one another... They land often in the vicinity of our cloisters. They teach us and reveal things to us which have been lost to us because of the cataclysms of the past centuries, which have changed the face of the earth."[7]

The same notion of the visitors from space, teaching us to live as brothers and sisters all, also features prominently in the accounts of the 1950s contactees, which generated such massive interest that governments became worried they would lose public support for their Cold War efforts. Several people have pointed out the role of the media in the efforts to confuse the public about the intentions of the space visitors, not only by peddling the official stories of 'weather balloons' and 'swamp gas', but also by sowing the seeds of fear and confusion through a steady supply of 'alien invasion' films. This was so obvious that the editors of *Flying Saucer Review* wrote in a special editorial in 1959: "We abhor this trend to condition world opinion through films and other media to fear the space ships."[8]

It should be acknowledged that George Adamski often reminded his audience of the cyclical unfoldment of creation and human progress. In prescient comments made in 1950, which seem even more pertinent for our time, he recognizes the difficulty that humans experience during the transition from one cycle to the next. This is because humanity, he says, "is bound so by habits and fears of the future that it resists the change, even if it takes its own life to do so – which could have the result of man's annihilation from this earth planet – it is a Divine plan and these characters that we sometimes consider so bad are nothing but actors to bring forth the result of this Divine plan. (…) But before we can have total consciousness of it, the old will have to be removed. And it is not easy to remove the old if the old has become a habit. In most cases, as we see it now, suffering or pain will be the medium through which this will come… For this has been promised – that at the end of 2,000 years from the birth of Christ, a new dispensation would be entered."[9]

Of course, the Christian churches have always held the promise of the Second Coming before their flocks. But equally, as Paul Brunton documents in his book *A Search in Secret India* in 1934, "All over the East there have been recurring hints of a coming event which is to prove the greatest thing history has given us for many hundred years. The prophecy of a Coming rears its head among the brown faces of India, the stocky people of Tibet, the almond-eyed masses of China, and the old grey-beards of Africa. To the vivid and devout Oriental imagination, the hour is ripe and our restless time bears outward portent of the near approach of this event."[10]

At the juncture in history between two cosmic cycles we witness the old structures failing, while the structures that will allow humanity to manifest the new cosmic influences are not yet in place. Increasingly, the more sensitive journalists and politicians are beginning to grasp this reality. "We live at a time in which the old is breaking down and the new does not yet exist," wrote columnist Rob de Wijk in the Dutch daily *Trouw* in September 2017. Less than a week later journalist Will Strong wrote about a plan from then-leader of the British Labour Party Jeremy Corbyn in *The Independent* newspaper, in which "business-ownership and income distribution are being rethought so as to fix a system that no longer functions...".

Long-standing dysfunction of political and economic systems always spawns conspiracy myths of secret groups ruling the world, like 'the Illuminati' or the 'deep state'. The latter was cleverly used by far-right activists posing as whistle-blowers ('QAnon'), fuelling public distrust by spreading rumours of paedophilia and other atrocities allegedly committed by the global elite. And, as always, people's uncertainty and insecurity that results when pandemics of fear, hatred and other viruses are fanned by intolerant and incompetent leaders who rely on an 'enemy' so they can take control of a system that is breaking down, are easily projected on 'the other' – whether it be the Jews in the 1930s or the refugees and migrants in our time.

Nevertheless, references in the media to "structures that no longer work" are becoming a regular occurrence and often leave the ether seem pregnant with a sense of urgency and the promise of a new world waiting to be born. Even 17-year-old climate activist Greta Thunberg spoke in these terms when she said, in June 2020, that society has reached a tipping point where injustice can no longer be ignored: "The climate and ecological crisis cannot be solved within today's political and economic systems. That isn't an opinion. That's a fact."[11]

Not surprisingly, this sense of anticipation is embedded

in the Ageless Wisdom teaching for the new age, but it is equally present in the information coming from space. In his contribution to *Crop Circles – Harbingers of World Change* archaeologist Michael Green outlines the connection between the UFO and the crop circle phenomena. He makes various references to the work of H.P. Blavatsky and Alice A. Bailey and concludes his chapter saying that various great cosmic powers have been released on humanity: "This process constitutes, I believe, the formal beginning of the New Age, or 'Second Coming' in Christian parlance. The crop formations are the visible sign to humanity that the Kingdom of God is upon us."[12]

In my previous books I quoted several contactees who had been informed likewise, for instance Enrique Barrios. The Age of Aquarius, he was told, is "a new evolutionary stage of the planet Earth, the end of millennia of barbarism, a New Age of love, a kind of 'maturation.' You have already entered the 'Age of Aquarius,' but only in time, not in deed. The Earth begins to be ruled by other kinds of laws and cosmic geological radiations. To put it another way, there is more love in people, but they continue to follow principles that have belonged to earlier, inferior levels of evolution. What occurs is a clash between what people feel internally and what they are obliged to do externally."[13]

Giorgio Dibitonto's space contacts told him unequivocally that this is a time like no other in history and that humanity would find at its side "a new Moses" who will "lead all the people on this new exodus, like a good brother or father".[14] Pierre Monnet, too, was told: "Throughout history spiritually advanced humans have been contacted by extraterrestrial beings to prepare humanity for the coming of an official messenger who taught man the rules and values to live in harmony with nature and the universe."[15] Earlier, Finnish American contactee Margit Mustapa referred to the return of a Teacher to inaugurate a new age in her first book, *Spaceship to the Unknown*: "The signs of [the] Aquarius age were predicting the coming of a new mankind. We are related to the world and its growth. What I am that will the world be... How can man stand still and be ignorant and awkward in his turnings and ponderings when the skies are full of visitors from other planets and are calling for a higher understanding of what universal outlook means in our present-day life."[16]

Indeed, as Vera Stanley Alder wrote: "Christ Himself told us that He would be with us at the end of the Age. It is now the end of the Piscean Age. Other Faiths also expect Him, under the names of the Messiah, the Maitreya and the Avatar. The disruption and stimulation everywhere in the world is due to the influence of the new Aquarian Age... Everywhere humanity is dividing into two camps – those who cling to the old ways, and those who rebel and seek for new ways."[17] And, "We can look upon the little child of today with awe and envy because he will live to see the greatest event of the new era – the 'second coming of Christ'. This event is foretold and expected by many thousands of people to occur before the end of this [20th] century."[18]

As we saw in chapter 3, under the Law of Cyclic Return the shorter evolutionary cycles are always inaugurated by a Teacher, a member of the kingdom consisting of those individuals who have evolved beyond the strictly human state. The Tibetan Master of Wisdom Djwhal Khul gives a brief but interesting overview of 'recent' appearances of such World Teachers which shows how the symbology in the resultant religions reflects the zodiacal signs of the cycle in which they appeared. In the age of Taurus, the Bull, "Mithra came as the world Teacher and instituted the Mysteries of Mithras with an (apparent) worship of the Bull. Next followed Aries the Ram, which saw the start of the Jewish Dispensation (...); during this cycle came the Buddha, Shri Krishna and Sankaracharya; finally we have the age of Pisces the Fishes, which brought to us the Christ."[19] According to Benjamin Creme, the World Teacher for the age of Leo was Hermes, for Cancer it was Hercules, and for Gemini it was Rama.[20]

The astronomical succession of cycles is explained on the NASA website as the 'precession of the equinox' and at the moment we are "again in transition, to the constellation of Aquarius, the water carrier. If you ever heard the song 'The dawning of the age of Aquarius' from the musical show 'Hair,' that is what it is all about."[21] The song, that is, not the new age. For what NASA wouldn't know – or subscribe to – is that each of such alignments brings with it specific energies from a cosmic source connected with the respective constellation in the Zodiac. The Aquarian age is "the age of synthesis, of inclusiveness and of understanding"[22], according to the Master D.K. and as it comes into power, humanity "will 'enter into' new states of awareness and into new realms or spheres of mental and spiritual consciousness, during the next two thousand years".[23]

In 1992, Benjamin Creme's Master wrote: "Whenever a new Light enters the world the effects are far-reaching, if not always immediately perceptible. Subtly, through the planes, that Light engenders new relationships, affecting the very nature of substance itself. Today, profound changes are taking place in the natural, as in the human world in response to the fiery forces now saturating this Earth's space and Being."[24] This time, the energies of synthesis which come in from the constellation of Aquarius must and will bring humanity to a universal manifestation of its innate oneness, while in the previous age of Aquarius, 26,000 years ago, it brought nomadic families to live in larger tribes and small nations.

As new energies gradually come in and saturate our planet, so the related ideas are slowly but steadily taking hold of the minds of a growing number of people. The social movements which the Master D.K. says were inspired by the Master Jesus in the 19th century, are a good example of the dawning sense of oneness, as shown in chapter 3. Likewise, the notion that our planet is not the only one bearing intelligent life has been seeping into the Western mind especially, even if it was never entirely lost among the more spiritually inclined tribes and races.

In 1926 the famous painter Nicholas Roerich, husband of Helena Roerich, who authored the Agni Yoga instalment of the Ageless Wisdom teaching between 1924 and 1938, was on an expedition in central Asia. In his travel diary Roerich describes an interesting episode that occurred while on the way back from Mongolia to Darjeeling: "On August fifth – something remarkable! We were in our camp in the Kukunor [Köke Nuur, or Upper Mongols] district not far from the Humboldt Chain. In the morning about half-past nine some of our caravaneers noticed a remarkably big black eagle flying above us. Seven of us began to watch this unusual bird. At this same moment another of our caravaneers remarked, 'There is something far above the bird.' And he shouted in his astonishment. We all saw, in a direction from north to south, something big and shiny reflecting the sun, like a huge oval moving at great speed. Crossing our camp this thing changed in its direction from south to southwest. And we saw how it disappeared in the intense blue sky. We even had time to take our field glasses and saw quite distinctly an oval."[25]

In chapter 2 we saw that in the 19th century the Masters of Wisdom already referred to the inhabitants of other planets. Working closely with the Masters, H.P. Blavatsky also said that the term human "does not apply merely to our terrestrial humanity, but to the mortals that inhabit any world, i.e., to those Intelligences that have reached the appropriate equilibrium between matter and spirit..."[26]

Returning to the reason for the increased presence of UFOs since the 1950s, it should be remembered that a new cosmic cycle does not start overnight. As the energetic influence of a new alignment gradually increases over a matter of centuries, the old energies fade out. The Masters of Wisdom began to prepare for the current age, which requires the emergence of the spiritual kingdom that is composed of these 'perfected' humans, into the full public view of the everyday world, in the 15th century.[27] The appearance of the Teacher also requires due preparation, especially if he does not come as before, by overshadowing the consciousness of a disciple, but Himself – this time in a world that, through transportation and the internet is now as physically connected as it is fundamentally One. This, by the way, illustrates the initial challenge for the Teacher – to bridge this psychological gap between our sense of ourselves as separate and the reality of our interdependent and interconnected relation with the world around us.

In April 1945 the Master D.K. wrote that in addition to "an upsurging of the Christ consciousness in the hearts of men everywhere" in connection with the advent of the World Teacher, or the return of the Christ in Christian terms, that through "overshadowing of disciples in all lands, He will duplicate Himself repeatedly. The effectiveness and the potency of the overshadowed disciple will be amazing."

In her second book, titled *Book of Brothers*, Margit Mustapa gives an extensive account of the teaching she received from one of the Space Brothers. About the present time she is told: "We are the Elder Brothers of your mankind. Receptive channels have to be opened from mankind. We are working with each one of you with a special materialization program, and the materialization of the spirit is our main task for the time being… Our mutual, main task is to materialize together the reappearance of the Christ… We are asking that each one of you, as well as the men of the world, live and think in a mutual love state and create stronger feelings, waiting for the Christ to come into the world once more and show us all the light, love and life! (…) The world-wide goodwill creates a new world full of Christs. Through it His second coming, expected at these momentous times, has happened!"[28]

More correctly, this is only one of three modes of return for the Teacher, according to the Master D.K. In addition to "the stimulation of the spiritual consciousness in man" as described by Mrs Mustapa and pronounced by e.g. Rudolf Steiner, D.K. says the second mode of return would be "the impressing of the minds of enlightened men everywhere by spiritual ideas embodying new truths, by the 'descent' (if I may so call it) of the new concepts which will govern human living and by the overshadowing of all world disciples". This we see reflected in the enormous strides taken in the scientific field toward embracing 'spiritual' realities. Thirdly, "we are told that Christ might come in Person and walk among men as He did before".[29]

Here we need to clarify who or what is meant by 'the Christ'. In the Christian doctrine the Christ is seen mostly as the "only begotten Son of God" or "God made flesh", who will return to punish the wicked and reward the faithful but, in the words of Benjamin Creme's Master, "No such being exists. In order to understand the true nature of the Christ it is necessary to see Him as one among equal Sons of God [i.e. all of humanity], each endowed with the full divine potential, differing only in the degree of manifestation of that divinity."[30]

As seen in the wisdom teachings, the Christ is not a denominational religious figure but rather an office in the spiritual hierarchy of Masters – in fact, He is the Master of all the Masters of Wisdom, the World Teacher who embodies the energy of Love, the Christ consciousness on our planet.

Appearing at the beginning of every cosmic cycle, a World Teacher usually manifests through overshadowing the consciousness of a disciple, in the way that the Buddha overshadowed Prince Gautama, or the Christ worked through Jesus of Nazareth. George Adamski also referred to this fact when he wrote: "Jesus, as an individual, schooled himself to permit [Christ's] consciousness to express through his form..."[31] Interestingly, describing the line of succession from the original Sankara (an earlier manifestation of Maitreya, the World Teacher), Paul Brunton is told that "the first Shankara promised his disciples that he would still abide with them in spirit, and that he would accomplish this by the mysterious process of 'overshadowing' his successors."[32]

One of the earliest experiments regarding the return of the Teacher was by the overshadowing of Jiddu Krishnamurti. This, according to D.K., was only partially successful because "The power used by Him [the Teacher] was distorted and misapplied by the devotee type of which the Theosophical Society is largely composed, and the experiment was brought to an end." D.K. added that "When Christ again seeks to overshadow His disciples, a different reaction will be looked for. (...) When Christ comes, there will be a flowering in great activity of His type of consciousness among men; when disciples are working under the recognition of the Christ, there will then come a time when He can again move among men in a public manner; He can be publicly recognised and thus do His work on the outer levels of living as well as upon the inner."[33]

According to Benjamin Creme, the experiment of preparing a possible vehicle for this manifestation of the World Teacher involved several other individuals as well, but when "Maitreya decided that He, Himself, would come" the plan "was changed and he [Krishnamurti] was no longer needed in that way."[34] Indeed, in his opening speech at the Star Camp in Ommen, the Netherlands, on 3 August 1929 Krishnamurti dissolved The Order of the

Star in the East. He left no doubt about the reasons for disbanding the organisation and rejecting the role of saviour that tens of thousands of followers around the world had cast him in: "You are all depending for your spirituality on someone else, for your happiness on someone else, for your enlightenment on someone else; and although you have been preparing for me for eighteen years, when I say all these things are unnecessary, when I say that you must put them all away and look within yourselves for the enlightenment, for the glory, for the purification, and for the incorruptibility of the self, not one of you is willing to do it. (…) For two years I have been thinking about this, slowly, carefully, patiently, and I have now decided to disband the Order, as I happen to be its Head. You can form other organizations and expect someone else. With that I am not concerned, nor with creating new cages, new decorations for those cages. My only concern is to set men absolutely, unconditionally free."[35]

According to followers of the Bulgarian mystic Petar Deunov, Krishnamurti stated that the world teacher "is in Bulgaria"[36], implying he referred to Deunov, but this statement is not found in his speech. Perhaps, if Krishnamurti said this at all, it was in a private comment to one of the 'devotee types' who were so numerous in the Theosophical Society and looking for an object for their devotion. And for all we know, Deunov may well have been one of the other individuals involved in the initial preparation experiment. Also, in his book *The Inner Life of Krishnamurti* adjunct professor of Philosophy and Religion at the American University in Washington, D.C. Miguel Angel Sanabria (writing as Aryel Sanat), documents that Krishnamurti never denied the existence of the Masters and Maitreya, who "were realities to Krishnamurti, apparently every single day of his life since he first encountered them in his youth".[37] Rather than denying their existence, he objected to people deifying or worshipping the Masters, or abdicating their own responsibility for the human condition.

The teachings of the Master D.K. through Alice A. Bailey, those of the Masters Morya and Koot Hoomi through Helena Roerich, and of Benjamin Creme's Master describe the various stages in an ongoing process of what will be a crucial event in the history of mankind and the evolution of this planet. One notable recurring characteristic in their writings is that they consistently emphasize the need for humanity's own involvement in its salvation, which requires individual recognition and practical application of the laws of Life.

Even in his earliest teaching, in the 1930s, George Adamski taught: "The path is a little different for each individual. Man himself must find himself, his duty as a form, his relationship to all forms and his oneness with the consciousness and intelligence which is manifesting through all. The vastness of knowledge and wisdom which can be attained by sincerity, honesty and love is unlimited, and the amount of service an individual can give is also unlimited. (…) It is well worth any effort that a man may make to steer his ship of life on the true course that leads to a vast concept of the interrelationship of all life."[38]

This was also impressed on the contactees. For instance, French contactee Pierre Monnet was told: "Everyone for themselves must become aware of the fact that every human being is responsible for all of humanity and that the terrible things that happen are the reflection of every individual's behaviour. For no single gesture, no single thought can ever be erased. Every thought, once sent out, creates hatred or love, happiness or unhappiness, life or death..."[39]

Likewise, a better, sustainable, more just and peaceful future must be built by humanity itself. In the second of the Agni Yoga volumes given by various Masters through Helena Roerich, the Master Morya writes: "By human hands must the Temple [i.e. the new civilization] be built."[40] Helena's husband Nicholas depicted the story that was introduced with this line in a painting called 'Signs of Christ' (1924), which hangs in the Roerich Museum in Moscow.

Despite the sizeable body of teachings given by the Masters since H.P. Blavatsky made the first attempt to inform the world about their existence, the majority of people still doubt the reality of this super-human kingdom in nature, if they are even aware of the notion. Perhaps it was to be expected that Blavatsky's desire "to bring the actual truth home to some who needed living *ideals* the most" met with mostly devotion and derision. Expressing her "passionate regret", she said: "That *which is not believed in, does not exist.* Arhats and Mahatmas having been declared by the majority of Western people as nonexistent, as a fabrication – do not exist for the unbelievers."[41]

And so, to the general public the idea of the Masters remained a fancy, while to the devotional type they became objects of worship. The same attitude caused George Adamski to regret his use of the term 'Master': "Because it has been so misunderstood and misinterpreted by students of certain philosophies, I have often regretted my use of the word 'Master' in referring to those wise ones whom I met on the space ships. (...) While these men denied perfection, their understanding of the Universal Laws was so advanced over mine, I naturally thought of them as masters in this respect. (...) Man's inability to recognize his potential has been the cause behind the perversion of our present teachings. He stands in awe of that which his comprehension cannot readily catalogue with fact already familiar to him, and these 'unknowns' he relegates to a realm of his own invention ... the mystical!"[42]

With contactees, religion, Masters of Wisdom, the new age, and Space Brothers all sorted, there remains only the pesky little question of time, or timing. If all of this is real, when will we have open contact? When will the UFOs land openly? When will the World Teacher and Masters of Wisdom step forward and put the stamp of authenticity on all this information?

In a statement from December 1919 describing the initial stages in the process of the coming of the Teacher, the Master D.K. said that the rate of rhythm of human consciousness "will be tremendously increased during the vital and unusual cycle upon which we entered in 1918, which will be tremendously speeded up in 1925, and will climax in a peculiar sense in 1934. We shall then enter upon three years of excessive endeavour in an effort to hasten the Coming and to prepare mankind for the great day of opportunity."[43] In 1922, he wrote that the coming of the World Teacher would take place "towards the middle or close of this present [20th] century".[44]

Apparently, these efforts were so successful, and the suffering of humanity during the war of such magnitude, that in 1947, in *The Reappearance of the Christ*, D.K. makes this momentous announcement: "The agony of the war, and the distress of the entire human family led Christ, in the year 1945, to come to a great decision..."[45] which "was reached during the period between the Full Moon of June, 1936, and the Full Moon of June, 1945 [and] resulted in the decision arrived at by the Christ to re-appear or return to visible Presence on Earth as soon as possible, and considerably earlier than had been planned."[46]

Benjamin Creme revealed it was hoped this would take place around 1950*, but humanity soon returned to its old divisive ways and the World Teacher postponed His return due to the Cold War. It was at the end of the same decade, in 1959, that Mr Creme was asked if he was willing to play a role in the preparation for the historic event of the coming of the Teacher for the new age when it would happen.[47] His mission, from 1974 until his death in 2016, was to inform the world of the fact that Maitreya had returned to the modern world in full physical presence, ready to make himself known as soon as humanity would be sufficiently responsive to his call for sharing and justice as the outer expression of the growing awareness of our oneness, our divine nature.

This event, according to Creme, will be broadcast around the globe on the Day of Declaration, during which Maitreya will address everyone over 14 years of age telepathically, so everyone will hear his message in their own language. While this takes place, humanity will

---

*See also the Introduction about the significance of this decade.

be given a profound experience of its oneness when "His energy, the energy of Love, will flow out in tremendous potency through the hearts of all."[48]

It is important here to remember that both D.K. and Benjamin Creme pointed out repeatedly that "His reappearance is conditioned and determined by the reaction of humanity itself; by that reaction He must abide. His work is subject also to certain phases of *spiritual and cyclic timing* and to impressions from sources to be found on higher levels than those upon which He normally works."[49] Of course, the same may be said for the reaction from the forces of materiality, and clearly these still have such a hold on humanity that thus far "...it has *not* been the will of man to establish right human relations. Hence the discipline of war, the torture of the forms, and the misery in human living today".[50]

Very much the same seems to apply to the appearance of the visitors from space, when George Adamski says: "Because the space travelers are human like ourselves their plans have to be changed or modified occasionally depending upon human reaction. It depends upon the people of this earth as to what will happen here in the next few months [written in January 1962]. As the future is subject to man's free will it cannot be predicted with great accuracy over short periods of time. Even though their space ships are being withdrawn, more and more people will land here from space craft and live among us as ordinary citizens. Even now they hold down regular jobs in laboratories of industry or governments. Many work in ordinary civilian occupations. In this manner they are fully informed of all plans and progress in all countries.

They are working in all nations in this manner, with the primary purpose of speeding man into space."[51]

However, when the visitors from space were met with secrecy and rejection many of George Adamski's forecasts, too, did not materialize within the time frame that he foresaw. For instance, in a talk in September 1955 he said: "You ladies, one of these days, will not have to be buying brooms. Your houses will probably be wired with magnetic lines of positive force around your floor, where your dust would be attracted. Since all particles are negative, the positive line would attract them like today a magnet attracts dusts of iron."[52] And, "We will find that our cooking, even the raising of food will be done electronically. On these ships they do have that. They have green vegetation by the day. We are now beginning to realize for the first time that the same force that propels our earth, produces life through the process of breeding which you and I depend upon. In other words, we are getting at the source of all things. Once knowing how to use this knowledge, nothing will be impossible. All things will be made quite easy. Man will live as long as he wants without knowing age."[53]

Given a similar task of creating a climate of 'hope and expectancy', in the early years of his mission Benjamin Creme stated that the earliest possible opportunity for the public emergence of Maitreya, the World Teacher was the late spring of 1982. When the emergence did not materialize for lack of attention from the world media – as the representatives of humanity's attention – many people, even among those who had their hopes up at a time of growing international tensions, decided Creme had just been another 'false prophet'. So why would we believe

Benjamin Creme's claims, for instance, that the World Teacher returned to the modern world in full physical presence in 1977, when no-one can point him out?

Creme himself always remained indifferent when people questioned him about the World Teacher not having stepped forward yet: "I have been told for over 30 years that He is emerging 'soon', but that is by the Masters' standards of time. They work in 2,000-year cycles, so a few years to the Masters is just like a Sunday afternoon. Maitreya's emergence is based on certain laws, above all the laws governing our free will. At any time since 1982 He could have been invoked by humanity performing just a few actions which would have allowed Him to come forward."[54] What's more, "As far as Maitreya is concerned, there is no set date, not even to appear on television. There are 'windows of opportunity'. These windows are constantly changing. They are the result of His understanding of the cosmic energies as they flow. These are positive and negative, and they change all the time."[55]

As noted above, the spiritual Hierarchy of Masters started preparing for this time no less than 500 years ago. It was then that the Masters saw the trend toward separatism in humanity intensify, according to the Master D.K., and a special conclave of all departments of the Hierarchy was called in around the year 1500 AD.

In preparation for the Age of Aquarius, "with its distinctive energies and its amazing opportunities", the Elder Brothers devised a plan for "the production of a subjective synthesis in humanity and of a telepathic interplay which will eventually annihilate time."[56] "To bring this about has been the objective of all training given during the past 400 years, and from this fact you can vision the utter patience of the Knowers of the race."[57]

This latter statement reminds us of what the Space Brothers told Italian contactee Giorgio Dibitonto about how they combat negative forces: "Our weapons are love, discretion, wisdom and patience. We are, nevertheless, far more effective in accomplishment than you could possibly imagine. (…) Soon your planet will comprehend this too, and the long awaited day will dawn for you too."[58] Hence, D.K. continues, the Masters "work slowly and with deliberation, free from any sense of speed, towards Their objective, but (…) They do have a time limit."[59]

One of the reasons that making accurate forecasts in our time frame is difficult for the Masters, according to Creme, is that they perceive reality "in the complete absence of time". They see the events of what we call the past alongside events which will take place in what we call the future. In a vision he was given by Maitreya in 1959 he saw events connected with the reappearance of the World Teacher: "I was inside a brilliant sphere of white light. The right-hand side of this sphere (there were no sides, but it felt like the right-hand side) showed in full colour, though there was a faint veil over the whole succession of scenes, all the events of the world which we call history… At the same time on the left-hand side, without turning my head and without losing awareness of the events of what we call the past, I was given a view, in colour, a panorama of the events which were going to take place, which we call the future – but which were *simultaneously* taking place… I was seeing that in the only time which exists, which is Now. And it is everlasting."[60]

Likewise, in answer to the question why the space

visitors asked Pierre Monnet not to be afraid of death, he replied: "Quite simply because they know that time does not exist and that the being that we identify with as the physical body, lived formerly, lives today and will live again tomorrow until its immortality in time and space. When you say: 'Life is short, I don't have time to achieve everything I want', you are mistaken. Because life is eternal. The time of a lifetime is not Time. The time of a lifetime is only a brief period of time in existence. It is, so to speak, nonexistent in relation to the infinity of eternity on the cosmic scale."[61]

A very similar notion of reality, or time, although explained in more 'mathematical' terms, comes from the space visitors who contacted Dutch businessman Ad Beers (writing as Stefan Denaerde). Explaining the difference between our everyday concept of time and 'infinite time', he relates the broader understanding of time as follows: "Our time is the *speed* and *direction*, from past to future, with which similar occurrences follow one another. This time can never be infinite. (…) Endless time [i.e. the Now] would have to have an infinite time speed and an infinite number of time directions from past to future. (…) Infinite time speed means that all events follow one another so quickly that they all happen at the same time."[62] This is why Benjamin Creme says: "Every event which is concerned with the emergence of Maitreya is already happening. The stock-exchange crash has already taken place. It is happening now because there is only Now. It is not waiting for a time. It is only we who are waiting for that time to precipitate on the physical plane as the event we call the stock-exchange crash. (…) And yet, to the Masters it is happening now."[63]

Interestingly, according to Steve Taylor in his book *Spiritual Science*, "The idea that time passes, or that there is a past and a future divided by a present, has never been verified by any physical experiments." He quotes Albert Einstein who said that "the distinction between past, present and future is only an illusion, even if a stubborn one". This, according to Taylor, "is also strongly suggested by some of the findings of quantum physics. In the microcosmic sub-atomic world, time seems to be indeterminate."[64]

In his text 'The Magnificent Perception', which shows all the hallmarks of having been inspired by a Master, George Adamski writes: "Time is the instrument used to measure the movement of Being – the element action creates in its path from the formless to the formed. In Eternity always you are, but in time you're unstable, inconstant."[65]

Perhaps this explanation from Mme Blavatsky will help our understanding further. Time, she says, "is only an illusion produced by the succession of our states of consciousness as we travel through Eternal Duration, and it does not exist where no consciousness exists in which the illusion can be produced, but 'lies asleep'. (…) The Present is only a mathematical line which divides that part of Eternal Duration which we call the Future from that part which we call the Past… the sensation we have of the actuality of the division of Time known as the Present comes from the blurring of the momentary glimpse, or succession of glimpses, of things that our senses give us, as those things pass from the region of ideals, which we call the Future, to the region of memories that we name the Past."

Recent research into how we experience the present confirms this "blurring of the momentary glimpse, or succession of glimpses". According to a study by the University of California's Knight Lab, "We focus in bursts, and between those bursts we have these periods of distractibility, that's when the brain seems to check in on the rest of the environment outside to see if there's something important going on elsewhere. These rhythms are affecting our behavior all the time." And although we experience a moment as continuous, "in reality, we've only sampled certain elements of the environment around us. It feels continuous because our brains have filled in the gaps for us", one of the researchers explains.[66]

Most people will have been aware of something similar on their daily run or commute to work, when we suddenly realize we "missed" a whole stretch of our route, not able to recall that we travelled it just minutes ago.

Blavatsky's explanation continues: "No one would say that a bar of metal dropped into the sea came into existence as it left the air, and ceased to exist as it entered the water, and that the bar itself consisted only of that cross-section thereof which at any given moment coincided with the mathematical plane that separates, and at the same time joins, the atmosphere and the ocean. Even so of persons and things, which, dropping out of the 'to be' into the 'has been', out of the Future into the Past – present momentarily to our senses a cross-section, as it were, of their total selves, as they pass through Time and Space (as Matter) on their way from one eternity to another: and these two eternities constitute that Duration in which alone anything has true existence, were our senses but able to cognize it."[67]

In the documentary *Infinite Potential* about quantum pioneer David Bohm, writer and director Paul Howard similarly says: "Everything we will come to know is already in formation, waiting to unfold into manifest reality. It is the implicate waiting to become explicate."[68]

The fact that time is an illusion also throws a new light on space travel, as explained by Benjamin Creme: "People think that if you are going to send something into the galaxy, it will take hundreds of years, the people would die before they could reach anywhere that was worth going to, and they would never return. It is not true. People come here from Mars and Venus in minutes."[69] This is because "time and space at a higher level do not exist, they are an illusion. Those who are already Masters of Space – the Space Brothers – can travel enormous 'distances', as we think of it, in seconds of 'time'."[70]

The Master D.K. stated that when we begin to sense the Plan of Evolution, "there comes the realisation of the unity of all beings, of the synthesis of world evolution and of the unity of the divine objective. All life and all forms are seen then in their true perspective; a right sense of values and of time then eventuates."[71] Because a Master is "completely unconditioned by Time as sensed in the three worlds of human experience", says Benjamin Creme's Master, he must make constant adjustments, "for instance, in order to accommodate His meaning and intentions to the state of consciousness of those still ensnared by Time." Hence the difficulty that the Masters have in 'translating' the events they foresee in terms of time as we experience it. In fact, the Master explains, the Masters do not see time "as an ongoing process, linking moments of action" but as "a state of mind". When we achieve a measure of oneness by aligning ourselves with our higher Self, we come to a more correct understanding of time, according to the Master, because this "is inherent in a correct relationship to our fellow men, for only when the sense of separateness no longer exists can a true realization of cyclic activity come about. A new world order, political and economic, is the essential prerequisite for this truer vision, for the required sense of Oneness can be achieved only when harmony and justice prevail. (...) From that new relationship between men will emerge the conditions on which a new sense of Time depends."[72]

When we realize the events that the Masters have been preparing for centuries will happen because they are already happening, Benjamin Creme says, "the more we will lose our impatience. We will find an ability to stand, to be there, and simply watch it take place on the outer physical plane..."[73]

About the timing of the World Teacher's declaration, Benjamin Creme's Master assures us: "Maitreya must make weighty decisions on slender – and changing – data. A fine line ... divides the necessary and the possible. Trust then the skill in action of the Lord of Love. Chafe not at the seeming delay of His appearance – in the all-embracing Now no such delay exists."[74]

Besides, according to Creme, the World Teacher and fourteen Masters of Wisdom are already present and inspiring change behind the scenes. Indeed, he says: "Maitreya's first political/economic effort, after coming to London in July 1977, was to inspire the Brandt Commission in November of that year"[75], whose 1980 report outlined the first steps towards establishing right human relations on a world scale. Likewise, Creme said, Maitreya visited Nelson Mandela while he was still imprisoned in Robben Island, forecast his release in July 1988, and inspired the peaceful transition of South Africa from apartheid to the multicultural majority rule that it is now.[76] He made many other forecasts that were published at the time and were subsequently proven accurate, like the changes in the Soviet Union[77], while the new cosmic energies led to the rise of people power, and the transition of dictatorships to democracies in South America in the 1990s.

At the same time, around the world, stories about a mysterious hitchhiker were reported who would tell the person picking them up that the Christ had returned, before mysteriously vanishing from the car. In 1991 therapist and counsellor G. Scott Sparrow published his collection of over 300 experiences of his clients who had told him of encounters with the Christ in his book *Witness to His Return*. Over the decades *Share International*, the magazine founded and edited by Benjamin Creme, has reported thousands of mysterious encounters and signs, many reported by its readers, but the majority sourced from media outlets around the world. All the while, various Jewish Rabbis are proclaiming that we may expect the coming of the Messiah any time now. Around 2016 leading Torah scholar Rabbi Chaim Kanievsky made headlines when he suddenly began to announce that the Messiah's arrival was imminent.[78] When Rabbi Shlomo Amar, the former Sephardic Chief Rabbi of Israel was about to leave for a trip abroad, Kanievsky asked him: "You are going abroad?

You don't know that the Messiah is standing at the door?" In August 2020 Rabbi Amar was quoted as saying: "All the great rabbis of this generation are saying that the Messiah is about to reveal himself. (...) All we need is to remain strong for a little bit longer."[79]

Until the time of their public emergence, the World Teacher and the Masters of Wisdom who have already taken their place in the modern world are working incognito, says Benjamin Creme, to allow people the freedom of thought to respond to their message and their work, and not to any presumed authority. The Space Brothers, too, respect our right to deny the obvious. Initially they contacted almost exclusively people with little to no formal education, lacking any 'authority' for their audience to look up to, and thereby leaving people free to respond to their message from the heart, to recognize the truth of their message from a heartfelt response. The fact that some of the original contactees later began to believe they were 'chosen' ambassadors or began to 'channel' spirits from space does not invalidate their original accounts, just as the talent of actors or musicians is not invalidated when they succumb to the pressures of mass adoration and begin to display antisocial or destructive behaviours.

Profound spiritual experiences may also cause the person involved to be temporarily unbalanced. This may happen, for instance, when we find our growth stimulated by the transfer of spiritual energy, a heightening of the rate of vibration of our composing bodies, and therefore an increased ability to absorb spiritual energies, through intermediaries like the Gurus and Avatars of the East.

More frequently such transfers happen more indirectly and less dramatically via disciples of the Masters who are sufficiently advanced to be 'impressed' and thus 'step down' the energies and ideas coming from the spiritual kingdom above the human, and even higher sources. These disciples may be active in the spiritual or artistic field, but in our time we may also find them in the political, economic, educational, and scientific arenas of human endeavour. Through their activity as a clearing house for higher energies they 'inspire' and lift up those around them and humanity as a whole to work towards the expression of the true, the beautiful and the good. This is also how art, for instance, may leave us feeling 'transformed', or how we are inspired to take up a humanitarian cause.

People who have attended *darshans*, meetings or talks with such disciples or avatars often attest to a feeling of 'heightened awareness' or experiencing such energies physically, emotionally, psychologically or even seeing them as a golden radiance, not seldom also captured in photographs. Listening to a talk by Krishnamurti, many of which have been made available on YouTube, often leaves one with a flash of insight, almost like a burst of energy or a sudden expansion of awareness, at the end.

A testimony from one of Dr MacDonald-Bayne's students describes similar experiences surrounding the talks he gave in Johannesburg, South Africa, in the spring of 1948 during which he was overshadowed by the Master Jesus: "It was not the actual words that made these lectures so amazing; it was the way they were delivered. The words themselves will ring in our memories for ever, but the greatest Truth was shown to us without words, and no description could ever convey the tremendous force of the Presence of the Master, and through him, the Love of the Father – a distinct evidence more real than anything physical around us." During the weekly lectures, the witness continues, "often the Master, quite visibly, takes over his body for an amazing moment, just at the end to give us his solemn blessing. The transfiguration is complete, with gown, beard and hair showing through the brilliant light around him..." Four other witnesses testified to very similar experiences.[80]

Similar experiences have also been described by people attending talks by Benjamin Creme, as well as George Adamski. When Emily Crewe attended Adamski's talk in Houldsworth Hall, Deansgate, Manchester on 1 May 1959, she found the hall packed to the doors "with a large audience sitting so quietly, you could hear a pin drop." In a description that will sound familiar to anyone who ever attended a talk by Benjamin Creme, she says how Adamski "liked to settle his audience before he began his lectures, to encourage a certain empathy with them, his large dark eyes searching the faces before him. Some intriguing stories were rife, about how some people were affected by this, by sudden, involuntary levitation or temporary loss of body weight. This happened to me, as, to my surprise and alarm, I felt myself lifted by a floating sensation of which I – with some embarrassment – immediately seized control by grabbing the wooden seat of my chair, which caused a further embarrassing clatter, with books and handbag hitting the floor."[81]

Through such transfers of higher frequency energy the cells in our being, and indeed our being itself, undergo a heightening of vibration, as such blessings literally 'inspire'

– blowing a breath of higher livingness into our being. And the measure in which we manifest the livingness that we have achieved through the process of evolution and our active attempts to reconnect with the Source through meditation or service to the Plan of Evolution, spurred on by blessings, is a reflection of our inner awareness of oneness and radiates outward into our surroundings.

This is what George Adamski refers to when he says that "each individual is a radiating center of influence, whose ultimate circumference no one can accurately perceive. (…) The world as a whole is composed of billions of individuals, each of whom is important as a radiating center of action. And the whole cannot be changed unless, and until, each small part is brought into cooperation and harmonious coordination with all others. In the human family, we know this as the Brotherhood of Man."[82] In this respect, Margit Mustapa was told: "You are to develop that selfless state of livingness, where the freedom of selfish I brings freedom to soul powers through the scientific knowledge of a radioactive human being"[83] – i.e. someone who radiates the quality of their soul, the energy of love.

A recent tragedy offers an example of how we experience radiating in sync with many others on the emotional plane. On 2 September 2015 3-year-old Aylan Shenu's body was photographed washed up on a beach in Turkey when his family, who had fled war-torn Syria and were hoping to join family in Canada, desperately tried to reach Greece in an inflatable boat at a cost of $5,860 for the four family members. Shockwaves of grief and horror travelled around the globe at the sight of the cruel loss of such an innocent life and the father's agony, and everyone in their right mind agreed such tragedies should never be allowed to happen again.[84] A similar global wave of synchronous response was felt in the wake of the killing of George Floyd, an unarmed black man who suffocated while being restrained with the knee of a Minneapolis police officer in his neck.[85]

Over the course of the days and weeks that follow, as the media attention is drawn to other news, we go back to our daily lives, refugees and migrants continue to drown in the Mediterranean Sea and black people continue to die at the hands of white police officers, because we live in a system that does not accommodate a sense of unity, but thrives on competition and separatism. And unless we see that this is how we enable the system that keeps us pitted against each other for a place under the sun that is our birthright, and unless we gather our wits and realize that our choices inevitably affect those around us and cause a ripple effect in society, even the physical and psychological environment, we can't and won't see the underlying reality that in a very real sense, 'they' are us, 'you' are me. And, therefore, unless and until we create humane living conditions for every single human being on Earth, we will not have true peace.

How, we may wonder, will we ever reach the stage where enough people may be convinced that only through a fundamental change in our view of life, that much coveted paradigm shift, will humanity survive today's culminating crises? When we look at recent developments and scientific studies about how change comes about, perhaps this is not as unlikely as it may seem at first sight. As artist Antony Gormley puts it in the documentary *Infinite Potential*: "Quantum physics invites us to be participators in that emerging of a world. And it has very fundamental, I think, both philosophical, spiritual and political implications, which are essentially that each of us is a co-producer of a world, that each of us is a co-producer of a possible future."[86]

When his space friends tell George Adamski that governments, pilots and scientists know about the extra-terrestrial presence on Earth, but "many have been muzzled and warned, and few dare speak out", he concludes: "Then it would seem that the answer lies largely with the ordinary man in the street, multiplied by his millions the world over." To which his Martian contact replies: "They would be your strength and if they would speak out against war in sufficient numbers everywhere, some leaders in different parts of your world would listen gladly."[87]

The extent to which we contribute towards the integrity of the planetary life and its kingdoms, seeing humanity as an integral part of the whole of this blue gem in the heavens, is in direct proportion to our ability to see and experience oneness inwardly. But how may this oneness be accomplished on a global scale when, according to Benjamin Creme, only about half the population, "the young make up the bulk of the incoming Aquarian people and the Piscean people make up the rest, which are all the governments of the world, all those who like the ways of the past, who have most of the money in the world"?[88]

Moreover, the 'Aquarian' half of the world population is not a homogenous group either. Social activists may frown on spiritual seekers for remaining aloof from real-world problems, while seekers may wonder why activists get wound up about the mere symptoms resulting from the spiritual causes behind such issues. Yet, their respective motives are not as far apart as they seem to the casual observer. A social activist's impulse to act stems from seeing how social injustice flouts the innate integrity of life, the oneness that she senses, even if she never articulated it. Likewise, a spiritual seeker responds to the heartfelt connection with the source of our shared humanity which he experiences, even if he struggles to make it practical. So, regardless of the angle from which we work to manifest the sense of oneness that triggers us, the common ground for our respective approaches is the fact that our growing consciousness seeks expression in right relations – with ourselves, with 'the other', and with the planet.

Returning to the definition of spirituality given earlier as the endeavour to overcome existing limitations, the activist may overcome her limiting condition by acknowledging the spiritual source of her impulse to act, while the seeker may do so by recognizing that an awareness of oneness needs to be acted upon in order to be brought into reality. Therefore, whether social activist or spiritual seeker, here too our efforts are better spent on seeking common ground than on emphasizing existing divisions.

Perhaps this will be easier knowing that religion and our own highest aspiration, systems science and esoteric wisdom teachings all acknowledge the fact of the evolution of consciousness as the impulse that drives progress in the manifestation of Life. Or, to repeat the words of systems philosopher Ervin Laszlo, "Through the action of love, an evolved consciousness fuses the elements of the mind into a higher unity. Attaining this 'spiritual evolution' is the meaning of existence."[89]

The individual effort to reconnect with the Source and manifest its qualities leads to the liberation of the spirit from the bondage of matter, which was symbolized in the Resurrection of Jesus in the Christian New Testament. The wisdom teachings tell us this 'resurrection' is not a one-time-only affair, but the goal for every individualized human soul. Every single being that is now a unit in the human kingdom will eventually reach that stage, because that is the end goal of the process of evolution on this planet. Or, in the words of professor Laszlo, "The closer the clusters [of coordinated vibration; i.e. physical forms] vibrate to the deep dimension, the more they are in-formed by the intelligence intrinsic to the cosmos."[90]

Given the present world situation, where the hard-won first steps toward social justice are under attack from the forces of commercialization who fear the loss of the obscene privileges which the status quo affords them, it is more important than ever that everyone, social activist and spiritual seeker alike, joins in this effort of "materializing the spirit" or "spiritualizing matter".

Significantly, in this respect, the Center for Strategic and International Studies reported in March 2020: "We are living in an age of global mass protests that are historically unprecedented in frequency, scope, and size. (…) The size and frequency of recent protests eclipse historical examples of eras of mass protest, such as the late-1960s, late-1980s, and early-1990s. Viewed in this broader context, the events of the Arab Spring were not an isolated phenomenon but rather an especially acute manifestation of a broadly increasing global trend."[91]

In fact, says Benjamin Creme's Master, "Many there are

today who, in their hearts, renounce the iniquities of the present materialism which pervades the planet. They long for justice and peace and march and demonstrate for their fulfilment. More and more, the peoples of the world are beginning to recognize that together they have the power to change the actions of powerful men. Thus does Maitreya trust the people and gives voice to their demands. Thus does He join their marches and adds His voice to theirs."[92]

And although protests sometimes escalate into violence or descend into plunder, two separate studies show that they should not be feared or shunned for such excesses. In one, social psychologist from Humboldt State University in California Amber Gaffney found that polarisation is a perfectly human response to our fallibilities and flaws. She says: "When people are highly uncertain of themselves and their motivations," [that is, when they are out of touch with their soul, the core of their being; GA] "different types of leadership become more attractive, like autocratic leaders in democratic societies" who play on this uncertainty using rhetoric like 'We're losing who we are'. Yet, some of the biggest positive social changes, she says, "have been the result of minority groups with strong cohesion and a clear identity. When we see positive social change, it comes from a minority. Think about the civil rights movement, women's vote. They are all incredibly positive, but they started with minority groups – they were the outsiders working against the norm."[93]

Compelling evidence for the feasibility of a better future is presented in a study by political scientist Erica Chenoweth at Harvard University, who found that "Countries in which there were nonviolent campaigns were about 10 times likelier to transition to democracies within a five-year period compared to countries in which there were violent campaigns – whether the campaigns succeeded or failed."[94] According to an article on the BBC Future website, Chenoweth's research "confirms that civil disobedience is not only the moral choice; it is also the most powerful way of shaping world politics…" But most importantly, perhaps, her study found that fundamental change does not require a vast majority of the population. Surprisingly, her research shows that nonviolent protests "engaging a threshold of 3.5% of the population *have never failed to bring about change*."[95] [emphasis added]

Working behind the scenes, the Masters have been training "a group of spiritually oriented men and women equipped to deal effectively with the problems of the time", according to Benjamin Creme's Master. "When the call sounds forth (…) they will take up the work for which they have been prepared: the reconstruction of our planetary life along entirely new lines. (…) Shortly, from the blueprint of the future now descending will precipitate the forms of the new civilization. Each nation has a part to play, bringing to the structure of the Whole its own particular voice. In this, the United Nations will play a vital role, co-ordinating the plans for reconstruction and redistribution."[96]

It is almost as if he refers to the work of such individuals when Dr Walach writes in the Galileo Commission Report: "During human history there were only few individuals who seemed to have had this gift of radical introspection, making the discovery of moral absolutes and bringing them into the cultural, political and religious arena of their time. In that sense it would neither be necessary nor likely that everyone needs to have the same experience and make the same discovery. It is also not necessary that everyone understands Einstein's equations and knows how to build aircraft. It is sufficient if a few people do and the others trust them."[97]

These are the initiates, individuals in the final stages of the evolution of consciousness into the next kingdom in nature, the spiritual kingdom of Masters of Wisdom. It is through such individuals, mostly unknown to the general public, that the cosmic energies from the constellation of Aquarius, now stronger than the energies from Pisces, will lead humanity to seek synthesis and unity. According to the Masters, they already have an immense impact, even if this has gone largely unnoticed by the media and the general public until now. But when we take a good look around, we see initiatives that would certainly qualify as contributing to the "reconstruction of our planetary life".

Here it is interesting to look at some developments that go mostly unreported, but may have a definite impact on how humanity will deal with the current crises. For instance, the World Economic Forum (WEF), which has made a name for itself as the impenetrable annual retreat for the global elite in Davos, Switzerland, recognized in its annual report for 2014 "that growing income inequality is an issue of central importance". Even more remarkable is that in June 2020, in response to the global Covid-19 crisis, the WEF launched The Great Reset initiative, "a commitment to jointly and urgently build the foundations of our economic and social system for a more fair, sustainable and resilient future" which "requires a new social contract centred on human dignity, social justice and where societal progress does not fall behind economic development."[98] It "will draw on thousands of young people in more than 400 cities around the world," called the Global Shapers Community, whose purpose is "to integrate young people as a strong voice for the future into global decision-making processes and to encourage their engagement in concrete projects that address social problems".[99] Interestingly, the Global Shapers Community was created by the same Klaus Schwab who founded WEF in 1971.

Whether the Great Reset will prove to be the initiative that will bring a sufficient number of people together to effect the required change on a global level, will depend in part on the extent to which the United Nations will be involved to ensure the engagement of the international community of nations in making the necessary decisions. Nevertheless, the initiative already has those worried who see freedom only in terms of governments abstaining from regulations that ensure socioeconomic justice, and can't see that true freedom begins with freedom from want and fear. Fearing the end of the world as they know and prefer it, they brand the Great Reset initiative as "social engineers revealing their hand" and a "Marxist agenda".[100]

However, according to the Master D.K., "It is interesting to note that the cycle now being inaugurated in the world is that of 'Growth through Sharing,' and that advanced humanity can now share the work, the responsibility and the trained reticence of the Hierarchy, whilst paralleling this and simultaneously, the mass of men are learning the lessons of economic sharing; and, my brothers, in this lies the sole hope of the world."[101]

Many who have been working for decades to spread the message of the Space Brothers, as others have for the Emergence of the World Teacher, may lament how long it is taking for their hopeful vision of the future to materialize. The parting words that major Hans Petersen wrote in his report of George Adamski's European tour in 1963 could have been written for our time: "And so you are alone – perhaps treading your own lonely path – in a world where contrasts meet more sharply than ever before. The masses drift hither and thither without plan, just following the doctrinal streams which the world's leaders spread through the newspapers, radio and television. Depression is great, as never before – murder, violence, crime, accident, swindle, sickness and stress, all to an extent so great that one cannot think that it could be worse. But the plan – which follows with the presence of the flying saucers in our atmosphere and the space people's work among us – will soon begin to be seen. When that happens a great many things will be changed; and then the pace will quicken. Progress will roll in over us, and the law of cause and effect will adjust the balance and equilibrium in our world."[102]

With our Elder Brothers, the Space Brothers urge us via Pierre Monnet: "We ask those among you who enjoy the privilege of truly understanding the law of love: do not lose patience, continue to show what love is and give unlimited love every minute of your lives, because before long the heart of man will change and you will be rewarded: the love that you have given will be returned a hundredfold."[103]

It seems the Masters know better what we are made of than we do ourselves. In the Agni Yoga volume

Supermundane – Book One, the Master Koot Hoomi writes in this respect: "We know the limitations of human ability, and We also know what can be expected of a human being in the building of a realistic future. We can just expect the highest degree of striving from Our messengers. When there is such intensity Our Magnet is active and serves as a strong shield. However, for the long journey timidity is not suitable. Everyone knows, in the depths of his heart, whether he is led by the highest degree of striving or is just being dragged along in fear. (…) Strong are those who are filled with gratitude, for their wings can grow! They will not be afraid of Our commissions. They know that We are greatly burdened, yet rejoice on the way to the Garden of Beauty!"[104]

With the real threat of fascism as the inevitable result of a relentless emphasis on material values – commercialization – and the planet itself at stake, these words from the Masters and – via Pierre Monnet – the Space Brothers take on a particular urgency at this juncture in human history, when we must take an active stand against policies that sow division and hatred: "We are here to help you, but that requires your unanimous consent. We hope that your hearts and minds will be filled with love and wisdom. We watch over you. But hurry, for the time is short…"[105]

Besides, the World Teacher himself has said in a message given through Benjamin Creme, we must not forget that "we are at the beginning and the end of a civilization, an epic period in the history of the world, and understand thereby that men feel the pain of change. For some it is a release into freedom. For others it is a loss of surety and calm. But, My brothers, the pain will be short-lived, and already many know this to be so. There is aid in abundance to help you through these difficult times. Accept eagerly this Age and recognize the signs of the new."[106]

When the World Teacher makes himself known, and our oneness in consciousness becomes the new paradigm for many of us, Benjamin Creme's Master says, mankind "will know that they are not alone in this vast universe. They will know that there are many other worlds in which their Brothers work for them, saving them from much harm. Maitreya will inaugurate the era of contact with these their far off Brothers, and will establish a future of mutual interaction and service. (…) He will show that the units of the one life manifest themselves throughout Cosmos; that until now this knowledge has been withheld from men but will provide a sure path for future generations to follow."[107]

As mentioned above, Benjamin Creme has always maintained that the World Teacher, who has been present and working in the world anonymously thus far, will present himself publicly by mentally overshadowing all of humanity in a globally televised event, on what he called the Day of Declaration.

In her autobiography Vera Stanley Alder shares a vision she was given in 1942 of "the greatest event in world history", of which her guide says: "...with all our history and research, we remain blind and asleep to the fact that this is indeed now the 'end of the Age' of which Christ spoke, and that therefore his Second Coming is closely upon us..." In the vision she was given, "We entered what appeared to be a large and beautiful college. People were streaming silently into it from all directions. We passed into a great hall at one end of which was an enormous television screen. The people assembled in utter silence, waiting. (...) A golden light gradually built up on the screen. It blazed with ever greater intensity until a figure suddenly appeared in its midst. I could hardly bear to look at Him. But He soon began to speak. His words were very few and very slow. They seemed to instil meaning and realization other than physically. I felt a great intercommunication at work in the great hall. Presently the figure faded out. The people remained silent and immobile, absorbing what had been given to them."[108]

In *Pioneers of Space*, which our inquiries suggest to be a largely factual account of George Adamski's out-of-body experiences with people from other planets, he describes an event that sounds very similar in its implication. Although clearly fictionalized, because his mission was not to be the standard-bearer for the new age, it is almost as if he couldn't help but share a profoundly inspiring experience he was given, perhaps comparable to the visions Mrs Alder or Benjamin Creme were given.

When the intrepid terrestrials manage to contact Earth from the Moon before their journey back home, they are asked to give an account of their experiences, for which the broadcast is shared among other radio stations, "so that all the people could listen in". During the broadcast a leading scientist from Mars extends his gratitude to the explorers from Earth: "...if all men of Earth were as these are at present, in the state of mind which has somewhat been changed to what it was when they left the Earth for this journey, your struggles and heartaches which you have been enduring through your mistakes in the planet Earth would cease to be, replaced with joys, happiness in a brotherly state and doing the will of Him, the Father who has placed you in one of His Mansions, the planet Earth."

Struggling, it seems, to find words to express the depth of the experience for humanity, Adamski continues his account: "And with this, the feeling of the people of the Earth reached an almost paralyzed state for they have never dreamed that anything like this could happen in their day. But it did, where the head scientist of Mars spoke to all the people of Earth. They felt as though they were paralyzed for it was like a voice from Heaven."[109]

The idea of humanity being 'overshadowed' in an almost celestial event that instils confidence and hope for the future, is not a very common element in storytelling. Yet, in his novel *The Amazing Mr Lutterworth*, Desmond Leslie describes a scene where the protagonist, Elias Minovsky, remembers his mission as a Brother from space when he arrives at the UN General Assembly, to inaugurate the 'Time of Splendour'. The first-person account relates: "All are now aware that something is about to happen. NOW. It is NOW. Out of nowhere I have appeared. They see me for the first time. They see and they hear me. (…) I begin by calming them at my sudden appearance … by embracing their minds in an overwhelming sensation of peace. That way I am able to attune them to me and to sustain us all as one mind, where they are assured that no magical trickery is employed. For the young man who has appeared out of nowhere is a man like themselves. And in this unity of mind I am able to speak to them not as nations, tribes and divided factions, but as men."

The Brother then proceeds to distribute the key to a new energy source, clean and free, among all the nations. This power, he tells the assembly, "shall change the face of the Earth. No more shall small groups, nor even single men, be able to rule multitudes through hunger in their bellies; for there shall be no more hunger nor want nor cold…"[110]

Earlier, in *Here to Help: UFOs and the Space Brothers*, I already pointed at the striking similarities between this fictitious account in Leslie's novel and Benjamin Creme's description of Maitreya's public emergence as the World Teacher on the Day of Declaration.

Given George Adamski's description of a similar scene of global import in *Pioneers of Space*, this episode in Leslie's novel could be evidence that Adamski was indeed given a similar vision of a future event as described by Benjamin Creme. Of course, in that case he would have been asked to not divulge it in public, so as to keep the focus on his own mission of broadening humanity's understanding of life to include the rest of the solar system. But he may well have shared his experience with intimates like his co-author Leslie who, after spending several months with Adamski and his group at Palomar Terraces in 1954, in turn may have used it in his 1958 novel, which Leslie later said was 75 percent non-fiction and based on Adamski's mission.[111]

Although the dawn of a new age and the coming of a new Teacher may not exist to the 'unbeliever', to paraphrase Mme Blavatsky, our investigation shows that the general acceptance of Earth being visited by people from elsewhere – and with it disclosure of what has been suppressed for so long – clearly depends on our acceptance and understanding of the growth of human consciousness, and manifesting this through the establishment of justice and freedom for all.

In all this, the role of the intermediary or messenger – be it Krishnamurti, Benjamin Creme, George Adamski or someone else, is to inspire hope and thereby breathe confidence into humanity to take responsibility for our individual growth, our own future. Taking this next step in our evolution means that we choose to speak up and actively contribute to creating correct relations with our fellow humans and the planet that gives us life, along the lines that have been indicated by our Teachers, the Brothers from space, and our own enlightened scientists. And unless we take action, with each of us individually adding our voice to the chorus demanding justice, freedom and peace for *all*, neither the Space Brothers nor humanity's Elder Brothers, not even the World Teacher himself can help us or do it for us. But when we do, we "will at once have countless brothers at our side", as Dino Kraspedon was told.

In the talks he gave in the UK on his 1959 lecture tour, George Adamski conveyed a special message from the Space Brothers, which already emphasized our crucial role: "Earthmen, you must find a better solution than war in which to sort out your economic problems. The way is now open for you to come out of your present treadmill of recurring frustration, pain and needless suffering caused by the regular altercations of war and peace and ignorance forced upon you by exploitation; exploitation by those of your brethren who do not want change and are happy with the world as it is. A window is now open through to the heavens, where your brethren out there wait for you to take the first steps to your true home among the stars…"[112]

References

1 Come Carpentier, 'Disclosure for India from America: IF Special with Stephen Bassett', 24 August 2020. See: <chintan.indiafoundation.in/podcasts/disclosure-for-india-from-america-if-specials-with-mr-stephen-bassett/>
2 Grant Cameron (2017), *Managing Magic. The Government's UFO Disclosure Plan*, p.275
3 George Adamski, letter to William T. Sherwood, April 1964. As quoted in Sherwood, 'UFO Understanding: An American Perspective', 17 July 1983
4 Benjamin Creme (1979), *The Reappearance of the Christ and the Masters of Wisdom*, p.206
5 Ibidem, pp.209-10
6 H.P. Blavatsky (1877), *Isis Unveiled*, Vol. I, p.627
7 Johann Gruber and Albert d'Orville (1672), *Voyage fait a la Chine en 1665*, as quoted in Michael Hesemann (1998), *UFOs The Secret History*, p.245
8 Special Editorial, 'Why This Horror From Space Trend?'. *Flying Saucer Review*, Vol.5, No.2, March-April 1959
9 Adamski, letter to Emma Martinelli, 13 March 1950
10 Paul Brunton (1934), *A Search in Secret India*, p.261
11 Jessica Murray, 'Tipping point: Greta Thunberg hails Black Lives Matter movement'. *The Guardian*, 20 June 2020
12 Alick Bartholomew (ed.; 1991), *Crop Circles – Harbingers of World Change*, p.142
13 Enrique Barrios (1989), *Ami, Child of the Stars*, pp.99-100
14 Giorgio Dibitonto (1990), *Angels in Starships*, pp.32-33
15 Pierre Monnet (1994), *Contacts d'Outre Espace*. Author's translation from the Dutch edition (1995), *Een boodschap van vrede*, p.125
16 Margit Mustapa (1960), *Spaceship to the Unknown*, p.123
17 Vera Stanley Alder (1976), *The Time is Now!*, p.3
18 Alder (1940), *The Fifth Dimension and The Future of Mankind*, p.124
19 Alice A. Bailey (1948), *The Reappearance of the Christ*, pp.126-27
20 Creme, 'Questions and answers', *Share International* magazine, Vol.21, No.3, April 2002, p.26
21 David P. Stern (2007), 'Astronomy of the Earth's Motion in Space – 7. Precession'. In: *From Stargazers to Starships*. See: <pwg.gsfc.nasa.gov/stargaze/Sprecess.htm>
22 Bailey (1954), *Education in the New Age*, p.3
23 Bailey (1948), op cit, pp.126-27
24 Creme's Master, 'The Fiery Light', June 1992. In: Creme (ed.; 2004) *A Master Speaks*, p.213
25 Nicholas Roerich (1929), *Altai-Himalaya. A Travel Diary*, pp.361-62
26 Blavatsky (1888), *The Secret Doctrine*, Vol.I, p.106 (6th Adyar ed., Vol.1, p.167)
27 Bailey (1955), *Discipleship in the New Age*, Vol.II, p.171
28 Mustapa (1963), *Book of Brothers*, p.152
29 Bailey (1948), op cit, pp.47-48
30 Creme's Master, 'The Son of Man', June 1984. In: Creme (ed.; 2004), op cit, pp.53-54
31 Adamski (1962), *My Trip to the Twelve Counsellors Meeting That Took Place on Saturn*, Part 2, p.1

32 Brunton (1934), op cit, p.124
33 Bailey (1955), op cit, pp.171-72
34 Creme (1979), op cit, p.203
35 J. Krishnamurti, 'Truth is a Pathless Land'. Speech at the Star Camp in Ommen, the Netherlands, 3 August 1929. See: <jkrishnamurti.org/about-dissolution-speech>
36 Boris Nikolov (1995), *Sunrise of The White Brotherhood*, Vol. 3, pp.77-79
37 Aryel Sanat (1999), *The Inner Life of Krishnamurti – Private Passion and Perennial Wisdom*, p.xiii
38 Adamski [n.d.; 1930s], 'Individual Analysis and Thought Control'. As reproduced in Gerard Aartsen (2019), *The Sea of Consciousness*, p.38
39 Monnet (1994). Author's translation from the Dutch edition (1995), op cit, pp.128-29
40 Helena Roerich (1925), *Leaves of Morya's Garden*, Book Two, p.75
41 Blavatsky (1973), *Collected Writings*, Vol.11, pp.292-93
42 Adamski (1957-58), *Cosmic Science* bulletin, Series No.1, Part No.3, Q60
43 Bailey (1957), *The Externalisation of Hierarchy*, p.518
44 Bailey (1922), *Initiation, Human and Solar*, p.61
45 Bailey (1948), op cit, pp.30-31
46 Ibid., p.69
47 Creme (1986), *Maitreya's Mission*, Vol. One, p.6
48 Creme (2007), *The World Teacher for All Humanity*, pp.15-16
49 Bailey (1948), op cit, pp.66-67
50 Ibid., p.113
51 C.A. Honey (ed.), 'Questions and answers', *Cosmic Science* newsletter Vol.1 No.1, January 1962, p.5
52 Adamski (1956), *World of Tomorrow*, p.2
53 Ibid., p.9
54 Creme (2010), *The Gathering of the Forces of Light – UFOs and Their Spiritual Mission*, p.83
55 Creme (2008), *The Awakening of Humanity*, pp.33-34
56 Bailey (1934), *A Treatise on White Magic*, p.403
57 Ibid., p.404
58 Dibitonto (1990), op cit, p.26
59 Bailey (1934), op cit, p.404
60 Creme (1997), *Maitreya's Mission*, Vol. Three, p.534
61 Interview with Pierre Monnet, *L'Ère Nouvelle*, January 2006. See: pointdereference.free.fr/m/www.erenouvelle.com/PORTCO-7.htm>. Author's translation from French.
62 Stefan Denaerde (1977), *Operation Survival Earth*, p.68
63 Creme (1997), op cit, p.540
64 Steve Taylor (2018), *Spiritual Science. Why science needs spirituality to make sense of the world*, pp.161-62
65 Adamski (1961), *Cosmic Philosophy*, p.5
66 Emma Betuel, 'Scientists reveal the number of times you're actually conscious each minute'. *Inverse*, 22 August 2018. See: <www.inverse.com/article/48300-why-is-it-hard-to-focus-research-humans>
67 Blavatsky (1888), op cit, Vol.I, p.37 (6th Adyar ed., Vol. 1, pp.110-11)

68 Paul Howard (dir.; 2020), *Infinite Potential. The Life & Ideas of David Bohm.*
   See:
69 Creme (2010), op cit, p.186
70 Ibid, p.59
71 Bailey (1944), *Discipleship in the New Age*, Vol. One, p.25
72 Creme's Master, 'A new age concept of Time', January 1982. In: Creme
   (ed.; 2004), op cit, pp.1-2
73 Creme (1997), op cit, p.541
74 Creme's Master, 'A Perennial Choice', July/August 1996. In: Creme (ed.;
   2004) op cit, pp.295-96
75 Creme (1993), *Maitreya's Mission*, Vol. Two, p.561
76 Patricia Pitchon, 'Maitreya's Perspective: The release of Nelson Mandela'.
   Share International magazine, Vol.7, No.7, September 1988, p.6
77 Pitchon, 'Maitreya's Perspective: The month ahead'. *Share International*
   magazine, Vo.7, No.6, July/August 1988, p.6
78 Adam Eliyahu Berkowitz, 'Leading Israeli Rabbi Says the Arrival of the
   Messiah is Near'. *Israel365News* [n.d.; 2020]. See:79 Ryan Jones, 'Top Rabbis: Look at the Signs, Messiah is Coming!'. *Israel
   Today*, 3 August 2020. See: <www.israeltoday.co.il/read/top-rabbis-look-at-
   the-signs-messiah-is-coming/>
80 Murdo MacDonald-Bayne [n.d., 1953], *Divine Healing of Mind and Body*,
   pp.9-16
81 Ragnvald A. Carlsen and Major Hans C. Petersen, 'Adamski – and a Feisty
   Lancashire Lass'. *Gensing Gardens News*, Vol.9 No.5, September-October
   2013, p.7
82 Adamski (1957-58), op cit, Part No.4, Q72
83 Mustapa (1963), op cit, p.144
84 'Death of Alan Kurdi'. Wikipedia. See: <en.wikipedia.org/wiki/Death_
   of_Alan_Kurdi>
85 'George Floyd protests'. Wikipedia. See: <en.wikipedia.org/wiki/George_
   Floyd_protests>
86 Howard (dir.; 2020), op cit
87 Adamski (1955), op cit, p.100
88 Creme, 'Questions and answers'. *Share International* magazine, Vol.35,
   No.9, November 2016, p.23
89 Ervin Laszlo (2017), *The Intelligence of the Cosmos*, p.41
90 Laszlo (2016), *What is Reality?*, p.15
91 Samuel Brannen, 'The Age of Mass Protests: Understanding an Escalating
   Global Trend. Center for Strategic and International Studies, 2 March
   2020. See: <www.csis.org/analysis/age-mass-protests-understanding-
   escalating-global-trend>
92 Creme's Master, 'The end of darkness', July/August 2005. In: Creme (ed.;
   2017), *A Master Speaks*, Vol. Two, p.43
93 William Park, 'How the views of a few can determine a country's fate'. BBC
   Future, 12 August 2019. See: <www.bbc.com/future/article/20190809-how-
   the-views-of-a-few-can-determine-the-fate-of-a-country>
94 Michelle Nicholasen, 'Nonviolent resistance proves potent weapon'.

The Harvard Gazette, 4 February 2019. See: <news.harvard.edu/gazette/story/2019/02/why-nonviolent-resistance-beats-violent-force-in-effecting-social-political-change/>

95 David Robson, 'How A Small Minority Can Change the World'. BBC Future, 14 May 2019. See: <www.bbc.com/future/article/20190513-it-only-takes-35-of-people-to-change-the-world>

96 Creme's Master, 'The emergence of great servers', April 1986. In: Creme (ed.; 2004), op cit, pp.89-90

97 Harald Walach (2019), *Beyond a Materialist World View. Towards an Expanded Science*, p.84

98 See: <www.weforum.org/great-reset/about>

99 See: <www.weforum.org/about/klaus-schwab>

100 Justin Haskins, 'Introducing "The Great Reset," world leaders' radical plan to transform the economy'. *The Hill*, 25 June 2020. See: <thehill.com/opinion/energy-environment/504499-introducing-the-great-reset-world-leaders-radical-plan-to>

101 Bailey (1955), op cit, pp.316-17

102 Petersen [n.d.], *Report from Europe 1963*, p.191

103 Monnet (1994). Author's translation from the Dutch edition (1995), op cit, p.123

104 Helena Roerich (1938), *Supermundane*, Book One, pp.222-23

105 Monnet, (1994). Author's translation from the Dutch edition (1995), op cit, p.131

106 Creme (ed.), 'Message from Maitreya'. *Share International* magazine, Vol.35, No.4, May 2016, p.3

107 Creme's Master, 'The way to the stars', April 2007. In: Creme (2017), op cit, pp.77-78

108 Alder (1979), *From the Mundane to the Magnificent. A Volume of Autobiography*, pp.198-99

109 Adamski (1949), op cit, p.254

110 Desmond Leslie (1958), *The Amazing Mr Lutterworth*, pp.199-200

111 Lou Zinnstag and Timothy Good (1983), *George Adamski – The Untold Story*, p.78

112 Carlsen and Petersen (2013), op cit

APPENDICES

I. The Ageless Wisdom teaching and the Space Brothers

According to the Ageless Wisdom teachings the history of humanity on this planet goes back 18.5 million years, and the evolution of consciousness has not stopped with the emergence of the human kingdom from the mineral, vegetable, and animal kingdoms in nature.

Over time, out of the human kingdom has evolved the spiritual kingdom of the initiates and Masters of Wisdom, those members of the human kingdom who have gone ahead in evolution and have completed the planetary Path of Return in consciousness or are in its final stages. Out of their midst, at the beginning of every new cosmic cycle or Age, a Teacher is sent into the world to reveal to humanity a further aspect of reality, of the Laws of Life, and our true spiritual Self, to guide and inspire humanity along the path of evolution to align our consciousness with the source of consciousness.

The body of knowledge that is now referred to as the Ageless Wisdom teachings is nearly 100,000 years old, and was gradually put together by the enlightened Masters of the time.[1] The teachings were first reintroduced to the modern world in H.P. Blavatsky's books *Isis Unveiled* (1877) and *The Secret Doctrine* (1888), which presented this new perspective on history and human evolution. Her seminal teachings have been elaborated in many publications, notably those of Alice A. Bailey, the Agni Yoga series, and Benjamin Creme.

H.P. Blavatsky says that the term human "does not apply merely to our terrestrial humanity, but to the mortals that inhabit any world, i.e., to those Intelligences that have reached the appropriate equilibrium between matter and spirit..."[2] As early as 1882 the Master Koot Hoomi referred to human and

244

super-human beings on other worlds when He spoke of "all the intelligences that were, are or ever will be whether on our string of man-bearing planets or on any part or portion of our solar system".[3]

The Master Djwhal Khul (D.K.) states that "in all the [planetary] schemes, on some globe in the scheme, human beings, or self-conscious units, are to be found. Conditions of life, environment and form may differ, but the Human Hierarchy works in all schemes."[4]

Benjamin Creme added: "All Hierarchies of all the planets are in touch with each other, and everything that takes place in an extraterrestrial sense takes place under Law."[5]

Through his associates, the World Teacher said in April 1989: "Humanity has naively believed that they are the only ones in space. But there are others there, far advanced, who have always watched over us, teaching us not to kill, to respect others, and to learn to be happy and free."[6]

To this end, according to Creme: "The Space Brothers have on this planet various people, like [George] Adamski and others, who are used to bring the reality of the Space Brothers to the world..."[7]

Adapted from: *Our Elder Brothers Return – A History in Books (1875–Present)*, www.biblioteca-ga.info.

References

1 Benjamin Creme, 'Questions and answers'. *Share International* magazine, Vol.33, No.7, September 2014, p.27
2 H.P. Blavatsky (1888), *The Secret Doctrine*, Vol.I, p.106 (6th Adyar ed. Vol.1, p.167)
3 A. Trevor Barker (1923), *The Mahatma Letters to A.P. Sinnet*, p.90
4 Alice A. Bailey (1925), *A Treatise on Cosmic Fire*, pp.358-59
5 Creme (2001), *The Great Approach – New Light and Life for Humanity*, p.129
6 Brian James, 'Maitreya's view – The Age of Light'. *Share International* magazine, Vol.8, No.4, May 1989, p.5
7 Creme (2010), *The Gathering of the Forces of Light*, p.36

II. The Golden Rule

The Law of Harmlessness,
as expressed in various religious traditions

Baha'i

"Lay not on any soul a load that you would not wish to
be laid upon you, and desire not for anyone the things you
would not desire for yourself." –*Baha'u'llah, Gleanings*

Buddhism

"Treat not others in ways that you yourself would find
hurtful." –*Gautama Buddha, Udana-Varga 5:18*

Confucianism

"Do not do to others what you do not want done to
yourself." –*Confucius, Analects 15.23*

Christianity

"Therefore all things whatsoever ye would that men should
do to you, do ye even so to them: for this is the law and the
prophets." –*Jesus, Matthew 7:12*

Hinduism

"One should never do that to another which one regards
as injurious to one's own self. This, in brief, is the rule of
Righteousness." –*Mahabharata, Anusasana Parva 113.8*

Islam

"Not one of you truly believes until you wish for your
brothers what you wish for yourself."
–*the Prophet Muhammad, Al-Muwatta, Hadith 13*

Jainism

"One should treat all creatures in the world as one would like to be treated." –*Mahavira, Sutrakritanga 1.11.33*

Judaism

"What is hateful to you, do not do to your neighbor. This is the whole Law; all the rest is commentary."
–*Hillel, Talmud, Shabbat 31a*

Native American

"Do not wrong or hate your neighbor. For it is not he who you wrong, but yourself." –*Pima proverb*

Sikhism

"I am a stranger to no one; and no one is a stranger to me. Indeed, I am a friend to all." –*Guru Granth Sahib, p.1299*

Taoism

"Regard your neighbour's gain as your own gain, and your neighbour's loss as your own loss." –*T'ai Shang Kan Ying P'ien, 213-218*

Wicca

"An it harm none, do what ye will." –*Wiccan Rede*

Zoroastrianism

"Whatever is disagreeable to yourself, do not do unto others."
–*Shayast-na-Shayast 13:29*

III. Universal Declaration of Human Rights

The Universal Declaration of Human Rights (UDHR) is a milestone document in the history of human rights. Drafted by representatives with different legal and cultural backgrounds from all regions of the world, the Declaration was proclaimed by the United Nations General Assembly in Paris on 10 December 1948 (General Assembly resolution 217 A) as a common standard of achievements for all peoples and all nations. It sets out, for the first time, fundamental human rights to be universally protected and it has been translated into over 500 languages. (*Source*: www.un.org/en/universal-declaration-human-rights/)

Human rights are rights inherent to all human beings, whatever our nationality, place of residence, sex, national or ethnic origin, colour, religion, language, or any other status. We are all equally entitled to our human rights without discrimination. These rights are all interrelated, interdependent and indivisible. (*Source*: www.ohchr.org/EN/Issues/Pages/WhatareHumanRights.aspx)

The selection of Articles on the next page shows that the UDHR not only offers protection against discrimination and oppression in the political, religious and social sense, but in the economic sense as well. The universal implementation of these Articles as the practical realization of the principle of sharing will create justice, provide unprecedented freedom for millions of destitute people, and signify the first step towards the outer expression of humanity's innate oneness.

For the full text of the UDHR, see:
www.un.org/en/universal-declaration-human-rights/

Article 22.

Everyone, as a member of society, has the right to social security and is entitled to realization, through national effort and international co-operation and in accordance with the organization and resources of each State, of the economic, social and cultural rights indispensable for his dignity and the free development of his personality.

Article 23.

(1) Everyone has the right to work, to free choice of employment, to just and favourable conditions of work and to protection against unemployment.

(2) Everyone, without any discrimination, has the right to equal pay for equal work.

(3) Everyone who works has the right to just and favourable remuneration ensuring for himself and his family an existence worthy of human dignity, and supplemented, if necessary, by other means of social protection.

(4) Everyone has the right to form and to join trade unions for the protection of his interests.

Article 24.

Everyone has the right to rest and leisure, including reasonable limitation of working hours and periodic holidays with pay.

Article 25.

(1) Everyone has the right to a standard of living adequate for the health and well-being of himself and of his family, including food, clothing, housing and medical care and necessary social services, and the right to security in the event of unemployment, sickness, disability, widowhood, old age or other lack of livelihood in circumstances beyond his control.

(2) Motherhood and childhood are entitled to special care and assistance. All children, whether born in or out of wedlock, shall enjoy the same social protection.

Sources and references

For references to the books by H.P. Blavatsky, Alice A. Bailey, Benjamin Creme, George Adamski and other works about the Ageless Wisdom teachings, contactees and related subjects, please consult the relevant sections at www.biblioteca-ga.info. Other sources are listed here.

Books

Richard Barrett, *The Evolutionary Human. How Darwin Got It Wrong*. Lulu Publishing, 2018.

Grant Cameron, *Managing Magic: The Government's UFO Disclosure Plan*. Itsallconnected Publishing, 2017.

Charles Darwin, *The Descent of Man*. John Murray, 1871.

Sarah Durston and Ton Baggerman, *The Universe, Life and Everything*. Amsterdam University Press, 2017.

Evelyn Fox Keller, *A Feeling for the Organism. The Life and Work of Barbara McClintock*. W. H. Freeman and Company, 1983.

Johann Gruber and Albert d'Orville, *Voyage fait a la Chine en 1665*. G Clousier, 1673.

Paola Leopizzi Harris, *Connecting the Dots. Making Sense of the UFO Phenomenon*. AuthorHouse, 2008.

Rey Hernandez, Jon Klimo, Rudy Schild (eds), *Beyond UFOs. The Science of Consciousness and Contact with Non Human Intelligence*, Volume 1. Foundation for Research into Extraterrestrial and Extraordinary Experiences, 2018.

Allen Hynek, *The UFO Experience: A Scientific Inquiry*. H. Regnery Company, 1972.

Alfie Kohn, *No Contest. The Case Against Competition*. Houghton Mifflin Company, 1992.

Ervin Laszlo, *The Intelligence of the Cosmos*. Inner Traditions, 2017

Ervin Laszlo, *What is Reality? The New Map of Cosmos and Consciousness*. SelectBooks Inc., 2016.

David Loye, *Rediscovering Darwin. The Rest of Darwin's Theory and Why We Need It Today*. Romanes Press, 2018.

Mohammed Mesbahi, *Towards a universal basic income for all humanity*. Matador, 2020.

Isaac Newton, *The Mathematical Principles of Natural Philosophy*, Book III: General Scholium. H.D. Symonds, 1803.

SOURCES AND REFERENCES

Diana W. Pasulka, *American Cosmic – UFOs, Religions, Technology*. Oxford University Press, 2019.

Nicholas Roerich, *Altai Himalaya*. Frederick A. Stokes Company, 1921

Wade Roush, *Extraterrestrials*. The MIT Press, 2020.

Steve Taylor, *Spiritual Science. Why science needs spirituality to make sense of the world*. Watkins, 2018.

Harald Walach, *Beyond a Materialist Worldview. Toward an Expanded Science*. Scientific and Medical Network, 2019.

World Wildlife Fund for Nature, *Covid-19, Urgent Call to Protect People and Nature*, 2020

Documentaries, essays, interviews, papers, lectures

Eben Alexander MD, 'My experience in coma'. 2012.

William Bains and Dirk Schulze-Makuch, 'The (Near) Inevitability of the Evolution of Complex, Macroscopic Life'. MDPI.com (Multidisciplinary Digital Publishing Institute), 30 June 2016.

Samuel Brannen, 'The Age of Mass Protests: Understanding an Escalating Global Trend. Center for Strategic and International Studies, 2 March 2020.

Silvano P. Colombano Ph.D., 'New Assumptions to Guide SETI Research'. NASA Technical Reports Server, 15 March 2018.

Gareth Cook, 'Does Consciousness Pervade The Universe? Interview with Philip Goff '. *Scientific American*, 14 January 2020

Paul Howard (dir.), *Infinite Potential. The Life & Ideas of David Bohm*, 2020

Dawna Jones, 'What is reality? Interview with Dr Ervin Laslzo', 28 November 2016

David Kipping, 'An objective Bayesian analysis of life's early start and our late arrival'. *Proceeding of the National Academy of Sciences*, 18 May 2020.

Klee Irwin, Marcelo Amaral and David Chester, 'The Self-Simulation Hypothesis Interpretation of Quantum Mechanics', MDPI.com (Multidisciplinary Digital Publishing Institute), 12 February 2020.

Xiameng Liu et al, 'Power generation from ambient humidity using protein nanowires'. *Nature*, 17 February 2020.

Auguste Meessen, 'Le Phénomène OVNI et le Problème des Méthodologies', *Revue Française de Parapsychologie*, Vol.1, No.2, 1998.

Michelle Nicholasen, 'Nonviolent resistance proves potent weapon'. *The Harvard Gazette*, 4 February 2019

Perimeter Institute for Theoretical Physics, 'Is the Universe a Bubble? – Physicists Work on the Multiverse Hypothesis'. SciTechDaily.com, 21 July 2014.

Martin Rees, 'Towards a post-human future'. The Royal Institution, 16 June 2020.

Rutgers University, 'No scientific proof that war is ingrained in human nature, according to study', 4 December 2018.

Russell Targ, 'Psychic Abilities'. SUE Speaks, 2013.

David Tong, 'Quantum Fields: The Real Building Blocks of the Universe'. The Royal Institution, 15 February 2017.

Alexander Wendt and Raymond Duvall, 'Sovereignty and the UFO', *Political Theory*, Vol.36, No.4, August 2008.

Tom Westby and Christopher J. Conselice, 'The Astrobiological Copernican Weak and Strong Limits for Intelligent Life'. *The Astrophysical Journal*, 15 June 2020.

Bob Wood, 'The secret relations between UFOs and Consciousness'. Society for Scientific Exploration, June 2019.

Websites

Brandt Equation: www.brandt21forum.info/

Climate Emergency EU: climateemergencyeu.org/

Darwin Project: www.thedarwinproject.com

Earth Overshoot Day: www.footprintnetwork.org/

Foundational Economy Collective: foundationaleconomy.com

George Adamski – The facts in context: www.the-adamski-case.nl

Global Shapers Community: www.globalshapers.org/

Great Reset Initiative: www.weforum.org/great-reset/about

Inequality.org: inequality.org/

Our Elder Brothers Return: www.biblioteca-ga.info

Peace in Space: peaceinspace.com

Share the World's Resources: www.sharing.org

Stop Ecocide: www.stopecocide.earth/

United Nations Organisation: www.un.org

UN High Commissioner on Human Rights: www.ohchr.org

Wellbeing Economy Alliance: wellbeingeconomy.org/

Index

INDEX

INDEX

263